COMPREHENSIVE BIOCHEMISTRY

ELSEVIER/NORTH-HOLLAND BIOMEDICAL PRESS

1 Molenwerf, P.O. Box 211, Amsterdam

ELSEVIER/NORTH-HOLLAND, INC.

52, Vanderbilt Avenue, New York, N.Y. 10017

First edition: 1977
Second printing: 1983

With 57 plates

Library of Congress Cataloging in Publication Data (Revised)

Florkin, Marcel, ed.
Comprehensive biochemistry.

Includes bibliographies.
CONTENTS.--section 1. Physico-chemical and organic aspects of biochemistry: v. 1. Atomic and molecular structure. v. 2. Organic and physical chemistry. v. 3. Methods for the study of molecules. [etc.]
1. Bilogical chemistry. I. Stotz, Elmer Henry, 1911- joint ed. II. Title.
QD415.F54 574.1'92 62-10359
ISBN 0-444-41544-0 (v. 32)

PRINTED IN THE NETHERLANDS

COMPREHENSIVE BIOCHEMISTRY

COMPREHENSIVE BIOCHEMISTRY

SECTION I (VOLUMES 1—4)
PHYSICO-CHEMICAL AND ORGANIC ASPECTS OF BIOCHEMISTRY

SECTION II (VOLUMES 5—11)
CHEMISTRY OF BIOLOGICAL COMPOUNDS

SECTION III (VOLUMES 12—16)
BIOCHEMICAL REACTION MECHANISMS

SECTION IV (VOLUMES 17—21)
METABOLISM

SECTION V (VOLUMES 22—29)
CHEMICAL BIOLOGY

SECTION VI (VOLUMES 30—34)
A HISTORY OF BIOCHEMISTRY

COMPREHENSIVE BIOCHEMISTRY

EDITED BY

MARCEL FLORKIN
Emeritus Professor of Biochemistry, University of Liège (Belgium)

AND

ELMER H. STOTZ

Professor of Biochemistry, University of Rochester, School of Medicine and Dentistry, Rochester, N.Y. (U.S.A.)

VOLUME 32

A HISTORY OF BIOCHEMISTRY

PART IV. EARLY STUDIES ON BIOSYNTHESIS

by

MARCEL FLORKIN

ELSEVIER SCIENTIFIC PUBLISHING COMPANY

AMSTERDAM · OXFORD · NEW YORK

GENERAL PREFACE

The Editors are keenly aware that the literature of Biochemistry is already very large, in fact so widespread that it is increasingly difficult to assemble the most pertinent material in a given area. Beyond the ordinary textbook the subject matter of the rapidly expanding knowledge of biochemistry is spread among innumerable journals, monographs, and series of reviews. The Editors believe that there is a real place for an advanced treatise in biochemistry which assembles the principal areas of the subject in a single set of books.

It would be ideal if an individual or a small group of biochemists could produce such an advanced treatise, and within the time to keep reasonably abreast of rapid advances, but this is at least difficult if not impossible. Instead, the Editors with the advice of the Advisory Board, have assembled what they consider the best possible sequence of chapters written by competent authors; they must take the responsibility for inevitable gaps of subject matter and duplication which may result from this procedure.

Most evident to the modern biochemist, apart from the body of knowledge of the chemistry and metabolism of biological substances, is the extent to which we must draw from recent concepts of physical and organic chemistry, and in turn project into the vast field of biology. Thus in the organization of Comprehensive Biochemistry, sections II, III and IV, Chemistry of Biological Compounds, Biochemical Reaction Mechanisms, and Metabolism may be considered classical biochemistry, while the first and fifth sections provide selected material on the origins and projections of the subject.

It is hoped that sub-division of the sections into bound volumes will not only be convenient, but will find favour among students concerned with specialized areas, and will permit easier future revisions of the individual volumes. Towards the latter end particularly, the Editors will welcome all comments in their effort to produce a useful and efficient source of biochemical knowledge.

M. Florkin
Liège/Rochester E.H. Stotz

PREFACE TO SECTION VI

(Volumes 30—34)

In the many chapters of previous sections of *Comprehensive Biochemistry* covering organic and physicochemical concepts (Section I), chemistry of the major constituents of living material (Section II), enzymology (Section III), metabolism (Section IV), and the molecular basis of biological concepts (Section V), authors have been necessarily restricted to the more recent developments of their topics. Any historical aspects were confined to recognition of events required for interpretation of the present status of their subjects. These latest developments are only insertions in a science which has had a prolonged history of development.

Section VI is intended to retrace the long process of evolution of the science of Biochemistry, framed in a conceptual background and in a manner not recorded in recent treatises. Part I of this section deals with Proto-biochemistry or with the discourses imagined concerning matter-of-life and forces-of-life before molecular aspects of life could be investigated. Part II concerns the transition between Proto-biochemistry and Biochemistry and retraces its main landmarks. In Part III the history of the identification of the sources of free energy in organisms is depicted. Part IV is devoted to early studies in Biosynthesis and Part V to the unravelling of biosynthetic pathways. While these latter parts are concerned with the molecular level of integration, Part VI is more specifically directed toward the history of molecular interpretations of physiological and biological concepts, and of the origins of the concept of life as the expression of a molecular order.

The *History* narrated in Section VI thus leads to the thresholds of the individual histories in the recent developments recorded by the authors of Section I—V of *Comprehensive Biochemistry*.

M. Florkin
Liège/Rochester E.H. Stotz

CONTENTS

VOLUME 32

A HISTORY OF BIOCHEMISTRY

Part IV. Early Studies on Biosynthesis

Chapter 41. From Vegetable Food to Animal Flesh

Chapter 42. Animals Recognized as Synthesizers

Chapter 43. Aspects, in the Field of Biosynthesis, of the Theory of "Protoplasm"

Chapter 44. Crystallization and Biosynthesis

Chapter 45. Biosynthesis Considered by Plant Chemists

I. Studies in Photosynthesis

II. Amino Acids Recognized as Precursors of Proteins and of Amides in an Appropriate Biological System (Plant Seedlings)

Chapter 46. Biogenetic Hypotheses Derived of the Known Behaviour of Plant Constituents

Chapter 47. Ureotelism and Uricotelism

Chapter 48. Ureogenesis

Chapter 49. Uricogenesis and Uricolysis

Chapter 50. Other Biosynthetic Aspects of Animal Chemistry

Chapter 51. Reversible Zymo-hydrolysis

Chapter 52. First Approaches to the Biosynthesis of Amino Acids

Section VI

A HISTORY OF BIOCHEMISTRY

LIST OF PLATES

(Unless otherwise stated, the portraits belong to the author's personal collection)

Volume 31

Errata and Corrigenda

p. 9, line 11, read *biochemistry*.
p. 19, ref. 4, instead of (*1969*), read (*1869*).
p. 97, line 3, instead of *myozymase extracts* [17-21], read *myozymase extracts* [18-21]
p. 124, line 14, instead of *arginine phosphate by acids*, read *arginine phosphate splitting by acids*.
p. 175, footnote, instead of *Coryell* [60], read "*Coryell* [60]

Introduction

One of the habits of historians consists in slicing up the succession of events in time into horizontal periods. Periodization involves the identification of a period which begins at one moment and ends at another.

It is easy enough, in the history of biochemistry, concerned with all molecular aspects of life, to identify theories introduced at a definite time and ending at the time of their demise, as is the case for biocolloidology, for the "cell theory" dealing with the alleged formation of cells in a blasteme, and for the theory of biochemical synthesis as a reversal of catabolic processes. These are biochemical theories which were accepted for a limited number of years. But, considering biochemistry at large, including its paleontology, periodization is in the present *History* considered as corresponding to the periodization of chemistry. Biochemical studies did not of course begin at the time of the emergence of biochemistry as a profession at the end of the 19th century. Alchemists, iatrochemists, pneumatic chemists and phlogistonic chemists have approached biochemical problems through their own conceptual system.

Parts IV and V of the present *History* are concerned with a biochemical theme: biosynthesis. The concept of biosynthesis as it appears in our present theory recognizes that the organisms build up their organic material from small chemical units. These units are "glued together by an enormously versatile condensing reagent, ATP" (Lipmann [1], p. 212). ATP is obtained by energy coupling mechanisms, either originating from redox potential or from conversion in group potential as was shown in Part III of this *History*. The concept of biosynthesis has replaced the concept of assimilation which was considered in Parts I and II.

References p. 10

From the point of view of the narration of events, Part IV begins with the entrance of Black in the Faculty of Edinburgh as a medical student, in the early 1750's. It ends with the introduction by Schoenheimer, in 1935, of deuterium in biochemical studies. It covers, from the point of view of general history, approximately two centuries, going from the end of the war of the Austrian succession to the eve of World War II. Its subject matter is the early period of studies on biochemical synthesis, which covers the period defined above. These early studies were performed before the introduction of the isotopic method, before the recognition of the energetic basis of biochemical syntheses and before the identification of synthases and synthetases.

As was stated in Part I, dealing with proto-biochemistry, the broad concept of "nutrition" has always involved a form of the concept of "composition" as opposed to "decomposition". We may recall the great importance, in the alchemist's "field of discourse" (Foucault [2]), of the concept of development based on the formation of a plant from a seed as well as the concept of order emerging from chaos as materialized in the Philosopher's Egg (see Chapter 2).

During the period of iatrochemistry, the famous experiment of Van Helmont (see Chapter 2) already represented a milestone in the approach towards biochemical synthesis.

But for the ancients, food, besides acting as a lubricant, brought a direct replacement of wear and tear, involving association of the classical four elements and eventually their transmutations. After the episode of iatromechanicism, the chemical approach to life was revived by the phlogiston theory. As was stated in Chapter 3, the first cycle of matter in Nature was formulated by the phlogistonists. Phlogiston, the principle of which was associated with the substance of all combustible matters, was considered as material. The theory was that food introduced phlogiston in animal organisms. Phlogiston was liberated by digestion. It left the body with the exhaled air (phlogistical air) from the lungs, becoming a constituent of the atmosphere, which was taken up by the plants and introduced in their combustible constituents. We find here a dual concept of Nature. The introduction of phlogiston within the framework of the constituents of plants is opposed to the liberation of phlogiston by animal digestion, with a production of animal heat. In this context, plants were, as phlogistoners, opposed to animals, considered as dephlogistoners. It has sometimes been claimed that, from a conceptual viewpoint, phlogiston can be identified with one of the four classical elements, fire. But fire or flame was free phlogiston, while fixed phlogiston was considered a constituent of all combustible materials, whether alive or not.

The first recognition of a chemical precursor for a biosynthetic process was the result of the interest in problems of a medical nature. As a consequence of his interest in a human disease, the stone, Black was led to recognize the production of "fixed air" in respiration; and it was a consequence of an interest in public hygiene, air pollution, which led the phlogistonists Ingen-Housz and Sénebier to recognize "fixed air" as the precursor of plant matter. The formulation by Sénebier of this process in terms of modern post-Lavoisierian chemistry followed on with the recognition of the composition of "fixed air" as expressed by the formula CO_2, as was taught by the new system introduced by Lavoisier.

One of the first teachings of the new chemistry was the persistence of chemical elements through material transformations. It was revealed that plant and animal materials are both principally composed of carbon, hydrogen, oxygen and nitrogen. This, as was stressed by Holmes [3], gave decisive support to the consideration of the phenomena of "nutrition" as chemical. While still maintaining this position, Holmes [4] qualifies his statement by the remark that

> "acceptance *in principle* that the ultimate level of material exchanges are chemical did not settle the question of what was the most appropriate level at which to investigate the phenomena — or in Foucault's terminology, where was the surface upon which the object of the study could be made to appear. This question itself can be looked at at two levels — whether the *effective* mode of analysis would be chemical or otherwise; and if chemical, at what level of chemical analysis" (p. 139).

During the period of modern biochemistry, starting with Lavoisier, a number of new concepts have successively been introduced in the conceptual system of biochemistry (Part II) and continue to be introduced. One of the major enrichments has been the identification, in organisms, of a collection of "proximate principles" ("immediate principles"), each characterized by an empirical formula and defined by properties (Chapter 5). Another concept dating from this period of inductions based on empirical formulas was the notion introduced by W. Prout that the nutrients as well as the

References p. 10

substance of organisms can generally be classified into three kinds of components, corresponding to our categories of carbohydrates, fats and proteins (see Chapter 5).

These concepts stimulated physiological chemists to endeavor to locate the emergence of "proximate principles" along the sequence classical with physiologists, of food → chyme → chyle → blood → tissues (see Part I). But this was a hopeless task, and in the 1830's the physiologists interested in biosynthesis turned their attention to the relation of the extreme points of the chain from plant food to animal flesh. In physiological research, the microscope remained one of the major instruments until the last third of the 19th century, when histology became an autonomous discipline. The recourse to primitive microscopes led to a belief in all kinds of artefacts or optical illusions, among which were globules considered to be present in food and intercalated in flesh. The intercalatory theory was part of the system of mechanical physiology introduced by Harvey and Descartes which considered an organism as a machine composed of levers, pipes, corpuscles, etc. within the frame of a corpuscular philosophy explaining Nature by the consideration of masses, figures and movements.

When the components of animal food were recognized as passing through capillary walls in the form of dissolved chemicals in the process of intestinal absorption, the globular theory was dismissed and attention was again focussed on the "proximate principles", identified by organic chemists, and on the relations of those "proximate principles" as they occur in food and in tissues. During the reign of the "cellular theory" (1838—1877), the "surface" of biosynthetic studies was shifted to the molecular level recently defined by chemistry. Molecules dissolved in the "blastemes" were considered to be involved in cell formation by molecular accretion. The shift coincided with the passage from mechanical physiology to the materialist biology promoted by Schwann, Helmholtz and Du Bois-Reymond in the 1840's, later by J. Loeb, and which was opposed to vitalism and to organicism.

Cell formation by an accretion of molecules (protoplasmic polymerization) remained for a long time one of the subjects of attention. Animal chemists and plant chemists were prone to speculate about the intermediates considered to be involved. Schwann had already compared the accretion of molecules in the successive layers he considered as formed around the nucleus, to a kind of crystallization. This is an analogy which, both before and after him, repeatedly returned in a variety of theories. Other hypothetical intermediates such as the protoplasm of Huxley, the living proteins of Pflüger and several variants of them, the "micelles" of biocolloidology, etc. have been suggested. We have, in Chapter 6, recalled a variety of postulated inframicroscopical "metastructures". Those trends led nowhere and finally, as shown in Chapter 16, the biochemists entered the cell in the steps of histologists and cytologists. Relying on a combination of differential centrifugation, electron microscopy and biochemical analysis, the localization of biochemical phenomena at the level of intracellular organelles led to our modern picture of cellular composition, activities and integration. Returning to the 1830's, at the time when organic chemistry was still understood as a chemistry of life, we find the chemists pursuing their collection and characterization of "proximate principles". Emphasis was directed to a comparison of the "proximate principles" of plants and those of animals. With the pressure of economic conditions, this relation became of primary interest for such outstanding organic chemists as Liebig, Dumas, Boussingault and Pelouze. They based their methodological approaches on the knowledge of the empirical formula of the chemical compounds and on the statical method, relying on balance sheet determinations. As was shown in Chapters 5 and 7, dealing with the general metabolic theories within the framework of changes introduced in the conceptual system of modern biochemistry, the chemists were accused of a lack of understanding of the processes actually taking place in organisms. If they were rightly charged, by the physiologists, of such shortsightedness, the chemists nevertheless contributed to a positive acquisition of knowledge by proposing the first chemical equations formulated to account for the mechanisms of the biosynthesis of natural substances. This positive contribution was particularly useful with respect to the occurrence of biosynthesis in animals, recognized after long and severe controversies within the camp of the organic chemists. They had the merit of maintaining a chemical theory of development and maintenance of organisms, while the biologists were led astray by a recurrence of organicist theories (such as the "organic creation" of Bernard). Organic chemists, unwarped by speculative theories, have retained and reinforced the doctrine forcefully exposed by Liebig, Dumas, Boussingault and others,

References p. 10

that biosynthetic processes were chemical phenomena. They consistently maintained the methodology of formulating a scheme of biosynthetic reactions, consistent with the knowledge of chemical structure, and repeating it in vitro. Such methodological requirements have become paramount.

After having separated from biology around the 1860's (see Chapter 12), organic chemistry, at the beginning of the present century, re-established a link with biochemistry through the work of such chemists as Emil Fischer, who proposed a new alliance,

> "as the great chemical secrets of life are only to be unveiled by cooperative work".

During the following decades this appeal was heard by a number of organic chemists, but many of them repeated the error of the organic chemists of the first decades of the 19th century and while dealing with chemical compounds present in organisms, they did, basing their interpretations on aspects of molecular structure, infer biosynthetic pathways again from the knowledge of the chemical behavior of organic compounds and not from reactions identified in the appropriate biological system. These hypothetical schemes of biosynthetic transformations

> "convinced chemists wholly, and biologists not at all" (Bu'Lock [5]).

Since around 1840, a number of physiologists had applied the vivisection method to biochemical problems. They took advantage of the contributions of organic chemistry which were increasing at the time. When, in the 1860's the organic chemists banned life from their field of studies, biochemical problems became the domain of physiological chemists and also of those scientists teaching chemistry in medical or agricultural schools.

Animal chemists and plant chemists have maintained their primary concern in the domain of chemical reactions identified in the appropriate biological systems. In the field of their experimentations on biosynthesis, the plant chemists, led astray by the formaldehyde theory of photosynthesis introduced by the chemists, failed to obtain experimental proof in its favor. On the other hand, studies on the biosynthesis of amides and of proteins in plant seedlings opened up new vistas. Animal chemists successfully explained a number of detoxication syntheses. They were able to untangle the skein of the excretion syntheses linked to the different forms of nitrogen metabolism, an aspect which pointed to the necessary definition and localization of the processes before they could be interpreted in chemical terms. The opposition between organic chemists on the one hand and plant or animal chemists on the other is not as clear cut as is sometimes suggested. Such organic chemists as Dumas or Boussingault were well informed on biological problems, while plant chemists such as Schulze or Prianishnikov, though working in agricultural institutions, were very competent organic chemists. Such leaders of animal chemistry as Baumann or Nencki had been trained in chemistry. The faculty of combining the identification of the appropriate biological systems and of unravelling the chemical reactions involved was naturally more frequent among this category of researchers.

The most impressive success in this field of experimental research was the unravelling of the biosynthesis of urea by the ornithine cycle, as revealed by the experiments of Krebs and Henseleit, accomplished on tissue slices. As most adequately stated by Fruton [6]:

"This achievement . . . marked a new stage in the development of biochemical thought. Not only was an explanation of a biochemical synthesis offered for the first time in terms of chemical reactions identified in the appropriate biological system, and not merely inferred by analogy to the known chemical behavior of the presumed reactants, but also the paper of Krebs and Henseleit provided a clue to the organization of metabolic pathways in living cells" (p. 436).

It took a few more years after the discovery of the ornithine cycle in 1932 before the nature of the cooperative work of chemists and biologists was considered in its proper perspective and before it was recognized, as was more generally the case after World War II, that the nature of biosynthetic pathways cannot be deduced from a mere consideration of the structure of the organic molecules. This requires recourse to experiments on the appropriate biochemical systems. Such appropriate experiments, aimed at unravelling pathways of biochemical synthesis as they take place in organisms, were rendered possible by such chemical and physical methods as the isotope tracer technique, the isolation of pure enzymes, etc. But the large body of knowledge about the chemical synthesis of organic compounds and about the chemical configuration of such compounds provided the indispensable epistemological background leading to the ways and means of application of different

References p. 10

strategies. These allowed for the unravelling of biosynthesis accomplished in vivo when, after 1935, the currents originating in organic chemistry as well as in plant chemistry and animal chemistry merged in the accelerated development which will be described in Part V.

As stated in Chapter 13, one of the leading concepts of modern biochemistry introduced in the course of its development within the framework of its conceptualist system was the notion of biocatalysis. Enzymology remained of only very limited importance during the 19th century. "Diastase" was identified by Payen and Persoz in 1833, pepsin by Schwann in 1836, and emulsin by Wöhler and Liebig in 1837. The enzyme concept was formulated by Berzelius in 1837.

But the physiological chemists remained reluctant during the second part of the 19th century with respect to the enzyme theory, and they were prone to prefer organismic views. We find, for instance, in a book of Neumeister [7] published in 1903, a statement of the lack of any demonstration of a biosynthetic process in a cell-free system, though since around the turn of the century the view had become current that the biochemical syntheses were a result of the reversal of catabolic processes, a concept which remained prevalent during the whole of the period considered here. During this period, since the early 1900's the enzymes assumed, as stated by Kohler [8], "the explanatory role that protoplasm had performed earlier".

Since Parts I and II (Vol. 30 of *Comprehensive Biochemistry)* appeared, the bookshelves of the history of biochemistry have been enriched with a series of important works. As several American critics have noted, a judgement with which the present author fully agrees, the book of Fruton [6], *Molecules and Life*, published in 1972, has been recognized as the best book available on the whole history of biochemistry. To complement this valuable contribution, Fruton [9] has published a very useful collection of references, under the title: *Selected Bibliography of Biographical Data for the History of Biochemistry since 1800.*

In his book *From Animal Chemistry to Biochemistry*, Coley [10] has described the approaches due to the animal chemists who have contributed to recognize the biochemical systems appropriate for the identification of chemical reactions. His book provides very useful information concerning the impingements of physiological, pathological or medical preoccupations on the progress of biochemistry.

Another contribution to the knowledge of the work of the animal chemists is the book by Holmes [11]: *Claude Bernard and Animal Chemistry. The Emergence of a Scientist.* This book covers the six years (1842—1848) during which Claude Bernard evolved into the position of a forefront leader. The book goes beyond the promises of its title and has a wealth of information on the whole picture of animal chemistry, and particularly on digestion and the source of sugar in animals. More than a hundred pages at the beginning of the book are devoted to the work of Dumas and Liebig in the controversial area concerning biosynthesis in animals. Chapter 42 of the present volume is based on this highly informative study.

In 1974 there appeared the book by Leicester [12], *Development of Biochemical Concepts from Ancient to Modern Time.* For Leicester, biochemical concepts refer to any hypotheses of bodily function which involve specific substances. He traces the development of these concepts from ancient to modern times.

The author expresses his thanks to colleagues and friends who have helped him by discussion and information, and particularly to H. Borsook, H. Chantrenne, A.C. Chibnall, H. Guerlac, H.A. Krebs, W. Niemierko, J. Rather and E.H. Stotz.

For permission to quote from copyright material, the author is indebted to J.S. Fruton, J. Parascandola and the American Academy of Arts and Sciences.

The author's gratefulness is also expressed to those who helped him to collect the illustrations. Their kind contribution is acknowledged in the list of plates.

References p. 10

REFERENCES

1 F. Lipmann, Wanderings of a Biochemist, New York, 1971.
2 M. Foucault, L'Archéologie du Savoir, Paris, 1969.
3 F.L. Holmes, Isis, 54 (1963) 50.
4 F.L. Holmes, J. Hist. Biol., 8 (1975) 135.
5 J.D. Bu'Lock, The Biosynthesis of Natural Products, London, 1965.
6 J.S Fruton, Molecules and Life, Historical Essays on the Interplay of Chemistry and Biology, New York, 1972.
7 R. Neumeister, Betrachtungen über das Wesen der Lebenserscheinungen, Jena, 1903.
8 R.E. Kohler, J. Hist. Biol., 8 (1975) 275.
9 J.S. Fruton, Selected Bibliography of Biographical Data for the History of Biochemistry since 1800, Philadelphia, 1974.
10 N.G. Coley, From Animal Chemistry to Biochemistry, Amersham, Bucks., 1973.
11 F.L. Holmes, Claude Bernard and Animal Chemistry. The Emergence of a Scientist, Cambridge, Mass., 1974.
12 H. Leicester, Development of Biochemical Concepts from Ancient to Modern Time, Cambridge, Mass., 1974.

Chapter 40

The Aerial Nutrition of Plants

1. Plants and the atmosphere

Comte considers chemistry as the science investigating the laws of the phenomena of composition and decomposition resulting from the molecular and specific interactions of the diverse natural or artificial substances. If we accept this definition, no biosynthetic pathway could exist before organic chemistry reached the stage of individualizing and characterizing biomolecules. On the other hand, if we define (Vol. 30, p. 1) biochemistry as covering all molecular approaches to biology, no biochemical approach to biosynthesis could have existed before biomolecules were defined. As we shall see (Chapter 42), the first tentative formulations of biosynthetic pathways were proposed during the 19th century. But if such pathways had not been individualized before, hidden as they were in the mass of the material studied, they were a constituent of this material and were indirectly considered in the context of a larger complex reality. In "nutrition", considered as covering the very complex sequence of events from food to organism, "transmutation", which had no chemical implication, remained, until Lavoisier, the current explanation.

One of the aspects of a chemical study of biosynthesis is the nature of the precursors involved. In the case of plant biosynthesis, it was during the period of phlogistonic chemistry that the concept of the "aerial nutrition" of plants was formulated and that "fixed air" was identified as a source of plant matter. Another concept also emerged at that time: atmosphere became considered as a link between plants and animals. This chemical progress must be considered as the first step in the unravelling of biosynthetic processes and, more particularly, in the recognition of the chemical precursors of the material constituents of organisms.

References p. 62

Plate 121. Stephen Hales.

2. Discovery of the fact that the life of an animal as well as the flame of a candle or of a lamp is extinguished by the removal of air

We have, in Chapter 3, mentioned the activities of those chemists of the 17th century who studied "airs", i.e. all kinds of elastic fluids.

It is in his first scientific book, *The Spring and Weight of the Air*, that Boyle, as early as 1660, showed that birds or mice die within 30 seconds to one minute of the time that air is removed by an air pump from the receiver. If the receiver is sealed tight but not exhausted of air, the animals die after 30 to 40 minutes (see Fulton [1]).

3. Black's "fixed air" and its physiological properties

Hales [2], in 1727, described examples of separation of an elastic fluid from several substances: animal, vegetable or fossil (i.e. inorganic substances found by excavating the earth). In such experiments, the loss of weight suffered by the material utilized was compensated by the weight of the "air" liberated, which was fixed in the material previous to its liberation. This elastic fluid was consequently called "fixed air" by Hales. This "fixed air" became the subject of Black's studies as a result of medical preoccupations: the search for a cure of stone, which has been defined as "one of man's oldest documented miseries" [3]. The cure of stone was the subject selected by Black for the thesis he presented at the Medical Faculty of Edinburgh. To quote Black himself:

> "I was particularly excited to it by the recent discoveries of the powers of lime water to give relief in cases of the stone and gravel, in which it was supposed to act by dissolving those concretions and expelling them out of the body" (cited after Guerlac [5], p. 137).

Why, in this context, did Black select "magnesia alba" (our magnesium sulfate) as the drug he studied with respect to its use against the stone? This point has been cleared up by Guerlac [5,6]

References p. 62

Plate 122. Joseph Black.

in his excellent biographical studies on Black. Guerlac recognizes the subject selected as linked to local aspects of life at the Edinburgh Faculty. In 1793, Hales had published the results of his attempts to cure the stone by dissolving gravel and calculi. The attempts had failed. Hales mentioned these vain trials in his *Statical Essays* [7].

When the famous nostrum of Mrs. Stephen * was bought by public funds (Coley [8]), a commission was appointed to report on its virtues. Hales was among its members. From clinical and chemical studies, Hales [9] concluded that the active ingredient of Mrs. Stephen's nostrum was lime. He also reported that he had succeeded in dissolving a few urinary calculi by lime in vitro. From 1745 onward, lime water was the medicine which attracted most interest with respect to the therapy of stone.

When Black arrived at Edinburgh, several professors were interested in the effect of lime water. A polemic was raging amongst them. Guerlac suggests that Cullen, who had long been interested in alkalis, may have called the attention of Black to the study of lime as a "lithontrypic" medicine and to its chemical properties. But why did Black, probably early in 1753, turn his attention to *magnesia alba* as a subject for his medical thesis, abandoning lime water? As shown by Guerlac [5], this change of orientation was in part due to the complexity of the problem. Another reason was the incongruity for a beginner to plunge into a polemic which was raging between professors of the Faculty. Black focussed his attention on a compound which had been distinguished from the other alkaline earths by the German chemist Hoffmann and was different from the calcareous ones. This powder, *magnesia alba*, had long been used as a mild purgative.

That Black had in mind a possible "lithontrypic" effect of the substance is clear, but this purpose was discouraged by the impossibility of preparing the equivalent of a lime water with magnesia. Black was fascinated with the discovery that there was an astonishing difference between *magnesia alba* and its calcined form *magnesia usta*. The first gave pronounced effervescence with acids and the second none. The *magnesia usta* did weigh considerably less

---

* Composed of an alkaline powder made by calcining egg-shells and garden snails, mixed with vegetable substances [8].

References p. 62

than the *magnesia alba* from which it was derived by calcination. What was the matter separated by fire and causing this loss of weight? In order to answer this query, Black submitted the *magnesia alba* to dry distillation. In the cooled receiver he found a small amount of watery fluid, corresponding only to one-sixth or one-seventh of the loss of weight observed in the *magnesia usta* when cooled. Remembering the experiments of Hales mentioned above, Black formed the hypothesis that the loss of weight resulted from a loss of an "air" which had escaped through the lute of sand and clay joining the receiver to the retort of his apparatus. This was consistent with one of the qualities of *magnesia usta*, its uniting with acids without effervescence.

"For", Black writes, "I began to suspect that the effervescence of the common magnesia proceeded from air [an elastic fluid] which it contained and which is expelled by the superior attraction of the acid; and that the reason why burnt magnesia did not effervesce was, that it did not contain this air, the air having been expelled from it by the action of heat".

Pursuing the analysis of his experimental data, Black recognized that the "fixed air" came, not from the *magnesia alba* but from an impurity, the "fixed alkali" (carbonate in our own language) used to precipitate the acid in the course of the preparation of *magnesia alba* *. What, through the detours of stone and of magnesia, was discovered by Black was finally the evolution of "fixed air" in the treatment of "fixed alkali" (our carbonate) by fire or by acids.

In his lectures, published in 1803 [4], but known and circulated in manuscript form after being delivered, Black records that he began in 1756, the year of the publication of his *Experiments* [10], to study the physiological properties of "fixed air". Impressed with the pertinence of the propositions of Van Helmont on *gas sylvestre*, Black confirmed that the gas given out in alcoholic fermentation was the same as the "fixed air" liberated by the action of acids on what we call carbonates, and making lime water turbid. Van Helmont had stated that the "air" produced by fermentation is the same as the *dundste*, or deadly vapor produced

* In the language of Black and his contemporaries, "volatile alkali" means ammonia, as distinct from "fixed alkali", of which two categories were known: "fossil alkali" corresponding to various impure forms of our sodium carbonate, and "vegetable alkali", or various impure forms of our potassium carbonate (also called *sal tartar*).

by burning charcoal. Black also put this conjecture to the test of lime water. Whatever his admiration for Van Helmont's contribution, and for instance for recognizing the similarities between the gas emitted in fermentation and the poisonous gas of the Grotto del Cane of Capri, Black was rightly convinced of the originality of his own contribution to the knowledge of "fixed gas" which had not been recognized by Van Helmont as produced by the action of acids on familiar "fossil substances" (our carbonates).

We have stated in Chapter 3 (p. 87) that it was Black who discovered in 1756 that "fixed air" was given up in respiration. To quote from his lectures [4]:

"I had discovered that this particular kind of air, attracted by alkaline substances, is deadly to all animals that breathe it by the mouth and nostrils together; but that if the nostrils were kept shut, I was led to think that it might be breathed with safety. I found, for example, that when sparrows died in it in ten or eleven seconds, they would live in it for three or four minutes when the nostrils were shut by melted suet. And I convinced myself, that the change produced on wholesome air by breathing it, consisted chiefly, if not solely, in the conversion of part of it into fixed air. For I found, that by blowing through a pipe into lime water, or a solution of caustic alkali, the lime was precipitated, and the alkali was rendered mild. I was partly led to these experiments by some observations of Dr. Hales, in which he says, that breathing through diaphragms of cloth dipped in alkaline solution, made the air last longer for the purpose of life" (quoted from Guerlac [5], p. 451).

Besides detecting the presence in expired air of "fixed air", Black formulated a metabolic notion and suggested that "wholesome air" ("atmospheric air"), or at least a part of it, is converted into "fixed air" in the course of respiration, and he recognized a similarity between respiration and combustion. Some of the discoveries of Black on "fixed air" reported in his lectures were the subject of confirmation or extension by his followers. In 1764, MacBride [11] (mentioned in Chapter 6, p. 129) published observations showing that the *gas vinorum* of Van Helmont, as well as the gas liberated in putrefaction, had the properties of Black's "fixed air". Cavendish [12] proved that only "fixed air" was produced in fermentation and that it was indistinguishable from the gas resulting from the action of acids on marble.

Black's "fixed air" (later called carbonic anhydride) is one of the varieties of the element "air" which were the subject matter of the experiments of the pneumatic chemists. From "atmospheric air" Priestley separated "dephlogisticated air", also called "vital air" or "fire air", because it is able to support life or flame (later called oxygen by Lavoisier). Cavendish obtained another "air" by passing atmospheric air over heated charcoal, characterized it as a chemical entity and distinguished it from "fixed air". The new "air" he called "mephitic air" (it was later called nitrogen). Cavendish communicated these results to Priestley, who published them in 1772 [3]. Independently of Cavendish, Scheele had, in 1771—1772, distinguished between "vitiated air" (synonym of "mephitic air") and "fire air" (synonym of "vital air") by exposing a confined volume of atmospheric air to various materials, such as moist iron filings or burning sulphur. Rutherford, as well as Cavendish, passed atmospheric air over heated charcoal and obtained "mephitic air". He was the first to publish this discovery in 1772.

Until the chemical revolution of Lavoisier, the chemical entities handled by the pneumatic chemists, in the field we are concerned with, were "fixed air", "mephitic air" and "vital air" ("dephlogisticated air"). Within the theory of phlogiston (see Chapter 3), which was generally accepted, a transmutation from one "air" to another was considered to be a consequence of changes in the proportion of phlogiston or in the modes of combination of phlogiston with "air". It must be remembered that the first chemical approach to biogenesis, in the case of photosynthesis, took place, as well as Black's approach to respiration, in this pneumatic and phlogistonic context.

4. Exchanges between plants and atmospheric air

Since the memorable experiment of Van Helmont (see Chapter 2, page 75) plants were considered as a product of a transmutation of water from the soil. That considerable quantities of air are inspired by plants was shown by Hales [2] in 1727. The background of this discovery is to be found in the new information which resulted from the introduction of a new scientific instrument, the microscope. One aspect of this new information was the discovery of stomata in the leaves of plants. Though the experiments of Hales made clear that exchanges took place between plants and atmosphere, the theory of the transmutation by the plant of the element water into the element earth appeared so simple that the addition of the element air was considered as trifle.

5. Priestley discovers that plants, contrary to animals, do not make the air noxious, and that plants restore noxious air

As has been stated in Chapter 3, the contribution of Priestley to understanding the relations of plants with air was set in the context of the phlogiston theory. Priestley always thought phlogiston; he was steeped in the theory and clung to it until he died. One of the main interests of Priestley was electricity and it is this interest which led him to study "mephitic air" (later called nitrogen), i.e. a kind of air which does not permit the life of an animal or the flame of a candle and which Cavendish had taught him to isolate from atmospheric air.

"Having read", he writes, "and finding by my own experiments, that a candle would not burn in air that had passed through a charcoal fire, or through the lungs of animals. . ." (quoted from Gibbs [15], p. 30),

Priestley tries to modify "mephitic air" by all kinds of means, including throwing

"a quantity of electric matter from the point of a conductor into it".

It was purely by chance that, during the summer of 1771, Priestley put a sprig of mint into a glass jar standing inverted on a vessel of water. After the sprig of mint had continued growing for some months, Priestley found that

"the air would neither extinguish a candle, nor was it at all inconvenient to a mouse which [he] put into it". . .

At the time it was commonly believed that not only animals but also plants rendered air noxious, and the vicinity of trees to houses was considered unwholesome. The result he obtained came as a surprise to Priestley who soon became aware of another surprising fact. Wondering what would result from putting a sprig of mint in air in which a candle had burned out, Priestley started the experiment on August 17, 1771 and on August 27 he observed that a candle burned in the air. He repeated the experiment with other plants, confirming the result. In his mind the result was to be explained by the withdrawal by the plant of an "effluvium" from vitiated atmosphere.

Benjamin Franklin's answer to the letter in which Priestley informed him of the results obtained, displays the prevailing

References p. 62

Plate 123. Joseph Priestley.

temper of American optimism:

"That the vegetable creation should restore the air which is spoiled by the animal part of it, looks like a rational system, and seems to be a piece with the rest. Thus fire purified water all the world over. It purifies it by distillation, when it raises it in vapours, and lets it fall in rain; and farther still by filtration, when, keeping it fluid, it suffers that rain to percolate the earth.

We knew before, that purified animal substances were converted into sweet vegetables, when mixed with the earth, and applied as manure; and now, it seems, that the same putrid substances, mixed with the air, have a similar effect. The strong thriving state of your mint in putrid air seems to shew that the air is mended by taking something from it, and not by adding to it. I hope this will give some check to the rage of destroying trees that grow near houses, which has accompanied our late improvements in gardening, from an opinion of their being unwholesome. I am certain, from long observation, that there is nothing unhealthy in the air of woods; for we Americans have everywhere in our country habitations in the midst of woods, and no people on earth enjoy better health, or are more prolific". (This letter is reproduced by Priestley [13] and is quoted after Nash [16].)

In the summer of 1772, Priestley repeated his experiments with sprigs of mint. On June 20 he allowed mice to die in a limited amount of air and he did put a sprig of mint in this air. A week later the sprig looked healthy and the air was restored, allowing a candle to burn and a mouse to breathe. At this time he had the opportunity of performing such experiments before two distinguished visitors, Benjamin Franklin and Sir John Pringle, who had just become President of the Royal Society.

For Priestley, what was taken up by plants from vitiated air, in his understanding, appeared as being phlogiston. For his contemporaries and for himself, animal respiration was essentially a process of taking up and releasing phlogiston, the color of blood changing in the process. This was in harmony with the statement in Plato's *Timaeus* (see Chapter 1) that substances coming from the stomach take up the color of fire. It was natural and logical to conclude, as Priestley did, that the "effluvium" taken from vitiated air by plants was phlogiston. This fitted in with the phlogistonic conceptual system. Phlogiston, liberated from the food by digestion, was carried by the blood and eliminated by the expired air, charged with phlogiston. The air phlogisticated by animal respiration was transformed by plants into "vital air" or "dephlogisticated air", one of the constituents of "atmospheric air", from which Priestley had isolated it. In this operation, the plants took

Plate 124. Sir John Pringle.

up phlogiston, purifying the noxious air and at the same time acquiring the quality of animal food, able to perform the phlogiston cycle of nature. In this scheme we already recognize the long lasting concept of the atmosphere considered as a link between plants and animals in the great cycle of Nature.

6. Pringle and the hygienic notion of the "goodness of air"

The medical impact of Sir John Pringle's career has been epitomized by Mrs. Singer [17] in the following terms:

"His view of medicine might almost be epitomized as cleanliness, to be achieved by purification of the air without and of the bloodstream within the patient. The first he would effect by ventilation, the second by correct diet".

Pringle [17] was the son of a Scottish baronet who was the neighbor of Lord Auchinleck, the father of James Boswell, Dr. Johnson's famous biographer. He studied medicine under Boerhaave in Leyden where he graduated in medicine in 1730. In 1742 an important event happened in Pringle's curriculum. He was introduced to John Dalrymple, the second Earl of Stair, who commanded the British army operating against the French in the Low Countries in the defense of Maria Theresa's claims to the Austrian succession. Lord Stair, who had also studied in Leyden, appointed Pringle as his physician. Pringle was soon raised to the position of "Physician to the British Army Hospital in Flanders", which carried a pension for life. When Lord Stair left command in 1744 he presented Pringle to his successor the Duke of Cumberland who appointed him as "Physician General to the Forces in the Low Countries and parts beyond the seas". With one interruption, when he was recalled to military duty in Scotland, Pringle stayed in Flanders until the end of the war of the Austrian succession and the signing of the Peace of Aachen in 1748.

Pringle, who had been appointed a member of the Royal Society in 1745, established himself as a practitioner in London. From his experience in the army, Pringle developed humanitarian preoccupations and, in 1750, he published a book which opened a campaign for public hygiene. A baronetcy was bestowed upon him in 1766.

References p. 62

7. First suggestion that "fixed air" is a food for plants

MacBride, who confirmed the production, mentioned by Black in his lectures, of "fixed air" in fermentation, as stated above, was convinced, as were other physicians, that "fixed air" could be used as a weapon against scurvy. How this therapeutic notion came to lead to a suggestion, however premature, of the role of CO_2 as a food for plants is a complicated chain of events. Priestley, who had for some time lived near a brewery (see Gibbs [15]), had been led by his natural curiosity and industry to realize that above the fermenting liquor in the vats existed a permanent supply of "fixed air" which Priestley used for the accomplishment of many an experiment. Among these was the observation that when a dish of water was placed in the gas it acquired an acidulous taste reminiscent of the mineral waters of Pyrmont or Selzer.

When in the spring of 1772, Priestley was dining with the Duke of Northumberland, the host showed to his guests a bottle of distilled water, which had been prepared from sea water by a naval surgeon, as suitable for providing seamen with a drink during long sea travels. Priestley, noting the unpleasant flatness of the beverage and thinking of the assumed virtue of "fixed air" suggested by MacBride and other physicians, remembered the experiments just recalled and suggested to the duke that "fixed air" be added to the distilled water prepared on board the ships by the distillation apparatus which was on trial at the time. The next day, Priestley got together a small apparatus for the preparation, for household use, of what became known later as soda water.

At this time, Captain Cook's vessels were preparing to undertake a trip, one of the purposes of which was to test foods reputed to prevent scurvy. Following the directions provided by Priestley, the Lords of the Admiralty, in May 1772, ordered the ships of Captain Cook, the *Resolution* and the *Adventure*, to be equipped not only with distilling apparatus but with the devices needed to carbonate the water (see Gibbs [15]). It was in the same line of thought that Priestley suggested to his friend Thomas Percival, of Birmingham, to study the possible medical uses of "fixed air".

Percival has acquired fame with his classical book *Medical Ethics* [18], first published in 1803 and still widely referred to today. * Percival settled in Manchester in 1767. With his friend, the apothecary Thomas Hardy, he founded the Manchester Literary and Philosophical Society, the weekly meetings of which were held at Percival's house.

In his *Observations on different kinds of air* (1772) [13] Priestley, on the basis of experiments accomplished in the gas above the fermenting liquor, had concluded that "fixed air" was fatal for vegetable life.

Priestley noted that a red rose, after being held over the fermenting liquor about 24 hours, became purple. Another turned white. Priestley suggested that his observations be checked with pure "fixed air" from chalk and oil of vitriol. This was tried by Henry, who reported his observations in the notes of his translation of Lavoisier's *Opuscules* [20]. He found the color unaffected.

Percival carried out other experiments with various plants and concluded that not only does "fixed air" retard the decay but

"actually continues the vegetation of plants, and affords them a *pabulum*, which is adequate to the support of life and vigour in them for a considerable length of time" [21] (p. 194).

Priestley did not take kindly to these ideas and opposed the thesis on the basis of the toxicity of "fixed air" for plants, in which they die. But as was later shown by de Saussure [22], plants thrive in concentrations up to 8 per cent, higher concentrations being toxic.

To Percival, Priestley wrote:

"In all these cases, you will say, I choke the plants with too great a quantity of *wholesome* nourishment; and to all yours I say, you do not give them enough of the *noxious matter* to kill them. Thus the amicable controversy must rest between us; and like other combatants, we shall both sing "Te Deum" (quoted by Gibbs, p. 132).

Later, in 1785, Percival wrote that he felt no inclination to exultation and preferred to drop the subject, noting that

"time . . . and the researches of philosophers [Sénebier] had thrown light on

* The text was made available for private circulation in 1794. Much of Percival's ethics was adopted by British and American professions in their ethical codes (Garrison and Morton [19], p. 154).

References p. 62

this disputed point" (cited by Scott [23] from *Manchester Memoirs*, 2 (1785) 326).

8. Priestley's green matter

In 1779, Priestley [24], informed that several researchers (for instance Scheele) had been unable to repeat his experiments, resumed them. In the happy innocence of his first experiments, Priestley had been favored by luck and also he had been performing his investigations in summer. He had no idea that light could play any part in the process of plants making the air unnoxious. Now experimenting in autumn and winter, he found it difficult to confirm his former results and his conviction was somewhat shaken. He reached the conclusion that irradiated water, by itself, liberates "dephlogisticated air", simultaneously depositing a non-organic green matter. On the other hand, he considered that plants may purify the air by dephlogistication.

The green matter in Priestley's experiments was not recognized by him as a plant, in spite of the opinion of some of his competent friends who, by looking at it under the microscope, recognized it as an alga (in fact, a kind of microscopical organism which was used extensively in modern research on photosynthesis). Furthermore, Priestley observed that the pump water, after being decanted, released gas bubbles. This liberation of "dephlogisticated air" dissolved in the water (a phenomenon little known at the time) he interpreted as a transmutation from water. This led him to consider that air could be purified by an irradiation of water, while in fact it was the result of an action on plant microorganisms contained in the water.

As Nash [16] remarks, this work of Priestley

> "presents a melange of the most ingenious and perceptive observations and experimentation combined with maddening failures in interpretation. Yet the "naked facts" were not reported in vain — they reached the attention of Ingen-Housz, who understood their import" (p. 368).

The epistemological obstacle which prevented Priestley from recognizing the vegetable nature of his "green matter" was mentioned by him in his book of 1781 [25] (published after the *Experiments on Vegetables* [26] by Ingen-Housz). As the green matter was produced in a close-stopped phial, Priestley was guided in his interpretation by his opposition to "spontaneous generation". He recognized in his book of 1781 [25] that:

"the seeds of this plant, which must float invisibly in the air, may have insinuated themselves"

through the cork stopping the phial.

9. Joint action of light and vegetation in the maintenance of the atmosphere (Ingen-Housz)

Ingen-Housz [17,27] was born in the Low Countries, in Breda, on September 7, 1749. His father, a leather merchant, is mentioned after 1755 as a pharmacist. During the war of the Austrian succession, British troops camped in the vicinity of Breda and the "Physician General" John Pringle became a friend of the Ingen-Housz family.

Jan Ingen-Housz graduated in medicine at the old University of Louvain (preferred by the Catholic minority of the Low Countries to which Ingen-Housz belonged) and settled as a doctor in Breda. He was to specialize, professionally, in the practice of "inoculation" against smallpox, in which he was to become one of the leading experts. On the other hand, he became a scientist according to the standards of the time, devoting his free time to experimentation, mainly concerning electricity, the gases of the atmosphere, and the thermal conductivity of metals. Before the introduction of vaccination by Jenner, inoculation was a useful weapon against smallpox. This method had first been brought to the attention of British doctors by letters addressed to the Royal Society by practitioners in Constantinople. In 1716, Lady Montagu had gone to Constantinople with her husband. She returned to England in 1718. At that time she introduced the practice of inoculation which became so popular that from 1776, at the Foundling Hospital, London, inoculation became mandatory.

Ingen-Housz, upon his father's death in 1764, received his share of his father's estate and began to spend much time first in Scotland and later in England where his protector Pringle introduced him to a number of outstanding personalities such as W. Hunter, J. Hunter, Priestley, Benjamin Franklin, etc.

References p. 62

Plate 125. Jan Ingen-Housz.

Ingen-Housz, at the Foundling Hospital of London, learned the Sutton method of inoculation (use of serum exuding from early smallpox pustules) and he soon became entrusted with all the inoculations by William Watson with whom he worked at the Hospital. He also developed a large inoculation practice.

When the Empress Maria Theresa and her son the Emperor Joseph II consulted their friend George III about the best choice of an inoculator to protect the latter's children from the terrible threat of smallpox, Ingen-Housz was, on Pringle's advice, sent (1768) to the Viennese court. His success was such that he was showered by the grateful parents with gifts and honors. He was appointed court physician with a life-long annual income of 500 gulden, which he later used to increase his independence in his career as an investigator. This career developed in parallel with his professional reputation as a competent inoculator, and gained for him the high recognition of being elected a member of the Royal Society. The first meeting he attended took place on March 21, 1771. At that time he remained in London until May 1772, to return to Vienna where he married the daughter of the botanist Jacquin and remained until 1789. During this period he occasionally travelled. As was mentioned in Chapter 3, one of these trips allowed him to attend, as a member, the meeting of the Royal Society during which the Copley medal was presented to Priestley. The eloquent speech delivered on that occasion by Pringle (Chapter 3, pp. 93–94) made a deep impression on Ingen-Housz and stimulated his imagination. In the course of his busy professional life, these thoughts developed in his mind and he conceived an outline of experiments which, during another trip to England, he developed during the months of June, July and August 1779 in a country house near London, and which he hurriedly had printed. This book was a milestone, as it demonstrated the action of light in the "dephlogistication" of air by plants (i.e. in the liberation of dephlogisticated air, later called oxygen). In the context of hygiene, the discovery of Priestley that noxious air was made unnoxious by plants was complemented by the notion of the limitation of the wholesome effect to periods of illumination. In the dark, just as animals did, plants contributed to make the air noxious. To this poetical contrast of plants between day and night (still persistent in popular folklore) Ingen-Housz added that flowers, those beautiful jewels of Nature,

References p. 62

"render the surrounding air highly noxious equally by night and by day". *

It must be noted that the tenor of the book is still a hygienic one, and that it still belongs to the era of phlogistonism and of pneumatic chemistry. It still was highly influenced by Pringle's preoccupations. Pringle introduced the practice of sea voyages as improving the health of "consumptive people", a concept which arose from his awareness of the harmfulness of marshy air, which had aroused his interest during his stay in Flanders. That Ingen-Housz remained highly stimulated by Pringle's train of thought is demonstrated by his publication in 1780 of a letter to Pringle in which he concludes as follows:

"It appears from these experiments that the air at sea and close to it is in general purer and better for animal life than the air on the land . . . so that we may . . . send out patients, labouring under consumptive disorders, to the sea, or at least to places . . . close to the sea which have no marshes in their near neighbourhood" [28].

But the essential contribution made by Ingen-Housz in his book of 1779 [26] is, as already stated in Chapter 3, the demonstration convincingly presented by him of the participation of light in the process of the liberation of "dephlogisticated air" (our oxygen) by plants. In his preface he quotes repeatedly from Priestley's book [24] published during the same year and there is little doubt that Ingen-Housz considered the interpretation by Priestley of the observations concerning his "green matter" as wrong, and that he was aware of the recognition, by several botanists, of this "green matter" as composed of green algae. On the other hand, Ingen-Housz owed to Priestley not only the broad esthetic and philosophical view of the complementarity of plants and animals, developed by Pringle in the field of hygiene, but he also got from Priestley the essentials of his chemical doctrine, the cycle of phlogistication and dephlogistication through plants and animals. Ingen-Housz recognized also the influence exerted by Bonnet on his train of thought.

Bonnet [29] had performed an experiment which stimulated

* Ingen-Housz writes: "I have seen school-masters so strongly prejudiced with this notion [that air emitted by young boys was fitter for respiration than that of old people] that they even would not allow the windows of the school to be opened, for fear that the young air, as they call it, of the schoolboys should escape; thinking that breathing this infectious and truly noxious evaporation would prolong their own life" (pp. 134—135).

the imagination of Ingen-Housz. Having noticed that bubbles of "air" cover the undersurface of the leaves of aquatic plants, Bonnet made the hypothesis that these bubbles are formed by air separated by the plant from the air contained in the water in which they lived. In order to put this hypothesis to an experimental test, he boiled water for three-quarters of an hour, to deprive it of dissolved air, and introduced into it a sprig of vine. No bubbles appeared, though the sun was warm. He then decided to reintroduce air to the water and for that purpose he blew through it, charging the water not with atmospheric air as he thought, but with pulmonary air, and the bubbles appeared again. Bonnet, believing that temperature influenced this phenomenon, noticed that the bubbles increased when the sun began to warm the water and disappeared, because of the cold, when night approached.

Though he missed the implications of introducing air by blowing, Ingen-Housz was astute enough to put to the test the alleged influence of sun heat as assumed by Bonnet. He thought that if heat was involved, the heat of a fire could produce the effect. Placing leaves in two jars of pump water and placing one in the sunlight outside and the other in the moderate heat of the vicinity of a fire inside, he recognized that the air obtained inside by the fire was very bad and that obtained outside in the light was dephlogisticated air.

Ingen-Housz was inclined to consider the production of dephlogisticated air by plants as a "transmutation" carried out under the influence of light. Nash [16] notes in this opinion an echo of Newton's sentiments quoted by Hales [2], in the following lines:

> "And may not light also, by freely entering the expanded surfaces of leaves and flowers, contribute much to ennobling the principles of vegetables? For Sir *Isaac Newton* puts it as a very probable query, "Are not gross bodies and light convertible into one another? and may not bodies receive much of their activity from the particles of light, which enter their composition? The change of light into bodies, and of bodies into light, is very conformable to the course of nature, which seems delighted with transmutation" (***Optiks****, query 30*) (quoted by Nash [16], p. 340).

By transmutation, Ingen-Housz imagines that the water itself or some substance in the water

> "is changed into this vegetation and undergoes, by the influence of the sun such a metamorphosis as to become what we call now dephlogisticated air".

References p. 62

Plate 126. Charles Bonnet.

This premonition of the origin of oxygen in the water was of course purely hypothetical. It was resumed later by Berthollet (see Nash [16]), but the final demonstration was only afforded in 1941 by Ruben, Randall, Kamen and Hyde [30]. These authors, using the heavy ^{18}O as a tracer, proved that the oxygen liberated in photosynthesis has its origin in water. Ingen-Housz considered that the transmutation of water (at the time still considered as one of the four classical elements) into dephlogisticated air (at the time considered as one of the elements, air, cleaned from phlogiston) "though wonderful to the eye of a philosopher" is not more astonishing than the change of grass into fat within the animal body or the production of oil in the fruit of the olive tree.

As we have repeatedly stated, at the time the concept of gaseous transmutability was widespread. Any gas (or air) was considered as derivable from another by a modification of its content in phlogiston, or of the degree of union with phlogiston. The concept of transmutation persisted until Lavoisier, and it was never actually disproved.

To quote Nash [16],

"it simply ceased to seem plausible and attractive after Lavoisier's new chemical system, with its doctrine of persistent chemical elements, had triumphed over the phlogiston theory" (p. 394).

10. The contributions of Sénebier

As we saw above, Percival's suggestion, according to which "fixed air" was a *pabulum* for plants, was rejected on the basis of its toxicity. It was Sénebier, fixing his versatile mind in the proper direction, who demonstrated that "fixed air" had an accelerating effect on the production of "pure air" by plants. This notion he expressed in a book [32] published in 1782, and confirmed in another book [33] the following year.

In these books Sénebier was starting out from a knowledge of the action of light on vegetables (Ingen-Housz) living in pump water (rich, as was known, in "fixed air") and from the current concept of the "purification of the atmosphere by plants" (Priestley).

Jean Sénebier [33—35] was born at Geneva in May 1742. Though inclined toward the natural sciences, he chose the career

Plate 127. Jean Sénebier.

of a minister, but his friends Trembley, Bonnet and Spallanzani * exerted a decisive influence on his work as a scientist. In 1773, Sénebier left his career as minister of the Church to become librarian of the Geneva Republic. He died as a result of surgery at Geneva in 1809. The reproach has been made to Sénebier of being a polygraph and dispersing his endeavors into many directions [35]. It is true that he had many interests such as literature, bibliography, physics, botany, methodology, philosophy, etc., but this catholicity was a general feature of the savants of the period.

As stated by Nash [16]:

"Sénebier was a meticulous and tireless experimentalist."

At the time of the publication of his books of 1782 [31] and 1783 [32], Sénebier was still living in a world ruled by "transmutations" and "phlogistications". "Fixed air" he considered a result of the phlogistication of the dephlogisticated air liberated by plants. In this primitive cycle (which is a cycle of phlogiston), the "fixed air" of the atmosphere is precipitated and dissolved in rain and ground water, from which it is taken up by the *roots* of plants. It provides phlogiston to plants which use it and release it in its dephlogisticated form. It would be an excess of presentism to understand Sénebier's views of 1782–1783 in the light of our modern views on CO_2. "Fixed air" was not for Sénebier at that time a combination of carbon and oxygen, but it was "an air" carrier of phlogiston, acting in an episode of the phlogiston cycle. In his book of 1783 [32], Sénebier recognized that a "vitiated air" could be improved by adding "dephlogisticated air".

We may recall here that it was Lavoisier who suggested that atmospheric air is composed of about one-quarter of oxygen (corresponding to dephlogisticated air, or pure air) and three-quarters of the inert gas nitrogen (corresponding to phlogisticated air, or mephitic air). Lavoisier introduced the concept according to which oxygen supports respiration (and combustion), in the course of which oxygen (pure air, vital air) is progressively replaced by CO_2 (fixed air) which is liberated by animals (Black) as well as by plants when they are in the dark (Ingen-Housz). The ancient con-

* Sénebier (Chapter 9, p. 165) published in 1807 the posthumous work of Spallanzani, concerning the observations of this great naturalist on the respiration of tissues.

cepts of phlogistication, vitiation or poisoning of the air by respiration were reduced by Lavoisier to oxygen deprivation and replacement of oxygen by CO_2 leaving a mixture (unable to support life) essentially composed of nitrogen and CO_2.

Though he was not, in 1782—1783, a disciple of Lavoisier's new theory, it appears that Sénebier, who postulated a "transmutation" of "fixed air" into "dephlogisticated air", was in this issue submitting to an indirect influence of Lavoisier's views. He reversed the transformation in respiration of oxygen into CO_2 (as supposed by Lavoisier) into a transmutation of fixed air into phlogisticated air, by plants (see Nash [16]) in photosynthesis.

Essentially, in his books of 1782 and 1783, Sénebier considers that the liberation of pure air by plants is the result of a dephlogistication, by the plants, of fixed air dissolved in ground water and absorbed through the roots. The role of the atmosphere, in this cycle, consists in a phlogistication of dephlogisticated air by the phlogiston of the atmosphere, the result being "fixed air", a variety of "phlogisticated air".

But the assumptions of Sénebier soon appeared indefensible. His contention that "phlogisticated air" was common air heavily charged with phlogiston did not resist the influence of Lavoisier's work. Sénebier believed that when the "dephlogisticated air" emitted by the plants was mixed with the "phlogisticated air" of the atmosphere, "fixed air" was formed by phlogistication. But Lavoisier was showing that the so-called "phlogisticated air" resulting from animal respiration was ordinary air in which a part of the oxygen had been replaced by an equivalent proportion of "fixed air". In a new book published in 1788 [36], and in a paper of 1792 [37], Sénebier's views present an epistemological evolution which has been aptly analyzed by Nash. This evolution led Sénebier from his former "gas transmutationism" and "phlogistonism" to the new concepts introduced by Lavoisier. This is attested by his use of the terms oxygen, carbon and carbon dioxide ("composed of oxygen and carbon").

Though still mainly concerned with hygienic preoccupations and still led by the concepts, centered on "pure air", introduced by Priestley and by Pringle, Sénebier, in his book of 1788 [36], writes in terms of modern chemistry. He considers that light can decompose, in the green parts of plants, the carbonic acid obtained from the soil through the roots (common knowledge at the time was that there was no carbon dioxide in the atmosphere). He considered that, in this process, oxygen was released from CO_2 and carbon retained, forming several "proximate principles" (oils, resins, etc.).

These views of Sénebier were in opposition to the theories of the school of Lavoisier (see Nash [16]) which had plunged into the matter with the work of Berthollet who suggested that illuminated green plants decompose water, releasing oxygen and retaining hydrogen. Another member of Lavoisier's circle, Hassenfratz, who was professor of physics at the Ecole Polytechnique, also considered that the action of light in photosynthesis consisted in a decomposition of water (a concept which was to prevail ultimately), and claimed that plants obtained carbon from the soil, as admitted by Sénebier, but not in the form of CO_2. For Hassenfratz the carbon of plants was obtained from other compounds collectively designated as coal (see below).

11. Conceptual evolution of Ingen-Housz

In 1789, Ingen-Housz published the second volume of his *Expériences sur les Végétaux* [38]. In this book, Ingen-Housz draws a picture of the concepts prevailing among the members of Lavoisier's school at the time. According to them, light did decompose water in the green parts, liberating oxygen which, united with "caloric" was liberated as "vital air". The hydrogen of the water, combining with the "coal" coming from the soil through the roots, formed carbonic acid, which enters in the composition of vegetable acids.

In 1796, Ingen-Housz published an article entitled *Essay on the Food of Plants and the Renovation of Soils.* * In this text, the accent is put on plant nutrition, more than on hygienic aspects.

> "I have deduced", he writes, "that of the two organized kingdoms, the animal and the vegetable, the animal derives its nourishment from the vegetable; but that the vegetable creation is independent of the animal world, provides for itself and derives its subsistence chiefly from the atmosphere. . ."

The link established between plants and animals is displaced

* This appeared as one of the appendices of the 15th Chapter of the *General Report from the Board of Agriculture*, London 1796.

Plate 128. Alexander von Humboldt.

toward metabolic aspects. This conceptual system of Ingen-Housz appears clearly in the following quotation:

"I discovered, in the summer of 1779, that all vegetables are incessantly occupied in decomposing the air in contact with them, changing a great portion of this in fixed air, now called carbonic acid. . . I found that the roots, flowers and fruits are incessantly employed in this kind of decomposition, even in the middle of the sunshine, but that the leaves and green stalks alone cease to perform this operation during the time the sun or an unshaded clear daylight shines upon them: during which time they throw out a considerable quantity of the finest vital air. . . I did not doubt that this continual decomposition of atmospheric air must have a general utility for the subsistence of the vegetable themselves, and that they derived principally their true food from this operation. . ."

Ingen-Housz goes on by stating that from the oxygen and carbon derived from the carbonic acid, the plants elaborate the substances they contain, together with the nitrogen he believes to be absorbed with the atmospheric air. In short, Ingen-Housz, in his paper of 1796 states that, from constituents of the atmosphere, the non-green portions of plants prepare carbonic acid which, carried to the green parts, in the light of the sun (as he had been previously demonstrated) is elaborated into oxygen (as Sénebier had suggested). This process is accompanied by a deposition of carbon in the green leaves and that carbon is available to the plants as a nutrient.

Nash [16] has aptly analyzed the source of these concepts. By adopting Sénebier's view that carbonic acid is decomposed in the green parts of plants, in sunlight, Ingen-Housz accounted for a source of carbon. Atmosphere should therefore constitute an adequate source of CO_2. But Lavoisier had concluded that air contained no carbonic acid. The way Ingen-Housz avoided this difficulty was to contend that the non-green portions manufactured fixed air from constituents of the atmosphere. On the other hand, the teleologically inclined Ingen-Housz liked the idea that the liberation of carbonic acid by plant respiration in darkness, far from being a "useless" vitiation of the atmosphere, was an essential nutritional operation of plants. It is in his paper of 1796 that Ingen-Housz states that carbonic acid must be considered as the "true natural food of plants".

It is obvious that the supposed absence of carbon dioxide in the atmospheric air was a source of great difficulties for Ingen-Housz. In 1798 a German translation of the pamphlet of Ingen-Housz

References p. 62

Plate 129. Nicolas Théodore de Saussure.

[39] appeared. In the introduction to this translation, von Humboldt mentioned analyses according to which the notion introduced by Lavoisier was incorrect and atmospheric air contained carbon dioxide (in concentrations of 0.7 to 1.4 per cent). We accept now that it contains less, 0.03 per cent in open country, but may contain much higher amounts near volcanoes, industrial sites, etc. (see Letts and Black [40]).

As stated by Rabinowitch [41], the concept of aerial nutrition opened a new period.

"After the appearance of Ingen-Housz's *Food of plants*, the problem of oxygen liberation by illuminated plants became merged with the problem of carbon assimilation from the air and the synthesis of organic matter" (p. 23).

This logical merging has been at the source of a number of theories which all constituted blind alleys until the participation of water was given its proper place (photosynthetic phosphorylation, see Chapter 38) and carbonic acid participation in the proper context of the carbonic acid cycle of reduction was acknowledged.

From the time of the publication of the pamphlet by Ingen-Housz in 1798, it was generally accepted that the biogenesis ("creation") of vegetables was independent of the animal world and that animals derived their nourishment from vegetables. From the work of the school of Lavoisier, and that of Sénebier and of Ingen-Housz, the interplay of chemistry and biology started within the broad field of biosynthesis.

12. The contribution of de Saussure

As stated above, Ingen-Housz was embarrassed with his inclusion of the carbonic anhydride of the atmosphere as a food for plants, by the assumed lack of this component in the atmosphere. When von Humboldt mentioned small amounts in the air, its scarcity in the atmosphere remained an obstacle to the recognition of CO_2 as the origin of plant carbon.

It was the great merit of the Swiss investigator Nicolas Théodore de Saussure [22] to demonstrate that the feeble but finite amounts of CO_2 present in the atmosphere are the source of most of the carbon of plants. Saussure was a member of a Geneva fam-

References p. 62

ily to which many eminent scientists have belonged. His father had invented the hygrometer. The method of Saussure was to determine the total carbon content of small plants growing in the open atmosphere while their roots were immersed in distilled water. If the CO_2 of the atmosphere was the source of the carbon fixed by the plants, these plants would grow more rapidly if more abundantly supplied with carbon dioxide. This was the case if CO_2 was added to air in concentrations up to 8 per cent. If this amount was exceeded, deleterious effects set in.

That plants achieve a substantial increase in weight when they grow in the open atmosphere with their roots in distilled water was, after all, only a more precise performance of the old experiments of van Helmont and of Boyle. But Saussure went further. He determined the carbon of the plants in comparison with a control: plants of the same age as those with which he had started his experiment; he observed that the plants which had continued growing in contact only with distilled water and atmosphere contained more carbon. As the concept of a "transmutation" of water had been excluded by the Chemical Revolution, there was no other conclusion than the one according to which the small amounts of CO_2 in the atmosphere are the origin of the carbon of plants. Saussure thereby gave a firm foundation to the concept of the aerial nutrition of plants which had been introduced by Ingen-Housz.

This concept was the subject of great excitement. Ingen-Housz already had gone as far as considering it in a passionate way and overlooking any other source of nutrient. But Saussure did not neglect the role of the constituents of the soil and of the water it contains. We shall relate his concepts on nitrogen sources in a following section. Saussure believed that the soil supplies nitrogenous materials and mineral materials to plants. He also reached the conclusion that water obtained from the soil is the major contribution to the weight of plants, providing hydrogen and oxygen. According to the views of Saussure, the chemical elements of which the substance of plants is formed originate from different sources: carbon from atmospheric CO_2, hydrogen and oxygen from water, and nitrogen from compounds contained in the soil. He considered on the other hand that the oxygen emitted by plants originates from CO_2 while the CO_2 emitted in the respiration of plants is derived from atmospheric oxygen. Saussure endeavoured to assign the source of supply of each of the major chemical elements of plants. This was the first complete scheme of a chemical conceptual system of plant nutrition. From a large number of experiments, Saussure summarizes his conclusions as follows:

. . ."Plants take up the hydrogen and oxygen of water, causing the latter to lose its liquid state. This assimilation is not very pronounced save when the plants simultaneously incorporate carbon. . . But in no case do plants decompose water directly, assimilating its hydrogen and eliminating its oxygen in a gaseous state. They emit oxygen gas only by the direct decomposition of carbonic acid gas. . . One cannot doubt that the greater part of the hydrogen that annual plants acquire during their development in the open atmosphere, supported by distilled water, has its origin in this liquid, which the plants "solidify". One can say as much of their oxygen. For one can judge, whether by the amount of carbonic acid gas that plants can decompose in a given time, or by the small change that they make in common air that the quantity of oxygen they secure from the atmospheric gases is entirely insufficient to account for the oxygen that they acquire in the short space of their development. It must not be forgotten that water is the most abundant product of the decomposition of most dry plants [opposed to succulent plants], nor that oxygen is their principal element" (translation by Nash [16]).

We presently are aware of the fact that the oxygen present in the CO_2 resulting from respiration originates from the organic compounds by the Krebs cycle (Chapter 33). We also know that the oxygen introduced appears in water and that the oxygen emitted by plants originates from water in the course of photosynthetic phosphorylation (Chapter 38). We also know that CO_2 is not decomposed by plants, but is fixed as such.

Saussure's conclusions were nevertheless carefully derived from his experimental results. He stated quite logically that growing plants draw upon the atmosphere for both carbonic acid and oxygen, and that they "metamorphose" the oxygen into carbonic acid and the carbonic acid into oxygen. But the latter process was difficult to reconcile with the extremely minute amounts of carbonic anhydride in the atmosphere.

As Nash [16] writes:

"However, illuminated plants carry out this metamorphosis with such astonishing efficiency that the vegetable conversion of carbonic acid to oxygen is far more extensive than the conversion of oxygen into carbonic acid. Consequently the net effect produced in the atmosphere by growing plants is the conversion of carbonic acid to oxygen" (p. 431).

References p. 62

Plate 130. Hugo von Mohl.

13. Opposition to the concept of the aerial nutrition of plants

In spite of the enthusiasm of Ingen-Housz and the careful experimentation of Saussure, the concept of the aerial nutrition of plants (carbon originating in atmospheric CO_2) was generally rejected until the forties. This can be attributed to a persistence of the old concept of the origin of plant nutrients in the soil, a concept which was supported by the school of Lavoisier, and for instance by Hassenfratz who, in 1792, had presented to the Académie Royale of Paris, three papers [42] in favor of the notion of the origin of plant carbon in the soil and the dung.

This concept persisted in the form of the so-called *humus* theory. The brown constituent of the soil, known as humus *, was considered to be the main source of the carbon utilized by plants. The humus theory was in flagrant contradiction to the classical experiment of van Helmont, while "aerial nutrition" accounted for the results of this justly famous experiment. It was only in 1840, in his influential book on agricultural chemistry, that Liebig [43] recalled the importance of the work on aerial nutrition by Ingen-Housz and Saussure (Sénebier, who was the first to recognize the importance of CO_2 as a food for plants, conferred this role on CO_2 of the soil, absorbed by the roots).

14. Photosynthesis and respiration

The concept of the continuous respiration of plants and their limitation of photosynthesizing to periods of illumination was difficult to establish. The continuity of plant respiration, even in light, was recognized by Ingen-Housz. It was Saussure who demonstrated that in the night the plants did produce CO_2, absorb oxygen and produce water. Saussure believed, without providing a demonstration of it, that in the light, simultaneously with photosynthesis, respiration was continued.

---

* The term *humus* was introduced by Diderot in 1765 (in the *Encyclopédie*) to designate the brown or black product of the decomposition of vegetable matters in soils.

References p. 62

In his book of 1851, von Mohl [44] did provide convincing data for the simultaneous performance of photosynthesis and respiration. The independence and simultaneous accomplishment of the two functions (which as we know now are respectively located in mitochondria and in chloroplasts) was generally recognized after the publication of the *Lehrbuch der Botanik* [45] of J. Sachs in 1868.

15. Atmospheric carbon dioxide recognized as the only source of the carbon of the organic matter of green plants

After the humus theory had fallen into disrepute, it became generally accepted that the carbon dioxide of the air was a source of the carbon of the organic matter of plants. That is was its only source was given general agreement due to the efforts of Liebig [43] and of Moll [46,47], and as a consequence the remnants of the humus theory were discarded. The problem then arose of recognizing the succession of products derived from carbon dioxide, from which the many products synthesized by plants are derived.

That the first products of the elaboration of carbon dioxide in green leaves were carbohydrates was hinted by von Mohl [44] and by Unger [48], but the first factual arguments in support of this view were afforded by Sachs [49]. Von Mohl [50] and Nägeli [51] had observed the presence of starch grains in the cells of green leaves and Sachs [49] studied the distribution of starch in the plant. He showed that in leaves kept in the dark for several days the starch grains disappeared while they were exposed to light. As a consequence of his experiments, Sachs designated the starch grains as the first visible products of assimilation of carbon dioxide.

16. Theories of the origin of plant nitrogen

Saltpeter (later recognized as potassium nitrate) had been known since the 14th century (literature in Aulie [52]). In 1658, Glauber had prepared it from animals, from plants and from soils. A little later, in 1661, Digby had increased plant growth by introducing saltpeter in soils. Saltpeter was commonly called "nitre" and it was the subject of a number of speculations (see Chapter 3). On the other hand, before nitrogen had been identified, ammonia had long since been known by the name of "volatile alkali". It was known that grass grows greener where the cows go and "volatile alkali" had been detected above dung heaps. In 1676, Edme Mariotte speculated on the presence of "volatile alkali" in the air and its falling with rain on the soil, penetrating the plant roots. It was Berthollet [53] who did show, in 1785, the presence of nitrogen (previously known by pneumatic chemists as "mephitic air") in an organic matter, plant gluten. He also [54] recognized ammonia (formerly called "volatile alkali") in the decomposition products of animals.

De Fourcroy [55] in 1789, showed the presence of nitrogen in the muscles of animals, carnivorous as well as herbivorous. This logically suggested a correlation: nitrogen of plants → nitrogen of herbivorous animals → nitrogen of carnivorous animals.

But where did the nitrogen of plants originate? According to Mariotte, it came from atmospheric ammonia. The rational deduction of the origin of plant nitrogen in air nitrogen, analogous to the penetration of atmospheric carbon dioxide in leaves, was formulated in a book by Chaptal [56] published in 1829. But during the first half of the 19th century, the two main current theories were the theory of the origin of plant nitrogen in manure introduced in the soil, and the theory of its origin in atmospheric ammonia (not in the atmospheric nitrogen).

17. Increase of soil nitrogen after legume cultivation

Empirically, the fertilizing virtues of legume cultivation had been known for a long time (literature in Aulie [52]). As early as the 1st century A.D., in his *Res rustica*, Columella gives directions to alternate beans and vetches with grass on a soil, in order to "fertilize" this soil. In Brabant and Flanders, there was an old tradition of crop rotation involving clover. Weston (1650) having visited these regions introduced clover in the English rotation practice. At this time (middle of the 17th century) clover was already

References p. 62

Plate 131. Claude Louis Berthollet.

in use in England but not in rotation. For the application of clover, sainfoin and lucerne in rotation, instructions were formulated by Tull in 1731. This empirical practice went over from England to France around the middle of the 18th century.

That legumes restore nitrogen to the land was demonstrated by Boussingault. Boussingault [52,57—59], born in Paris in 1802, had, at the age of 18 years, entered the Ecole des Mines of Saint-Etienne from which he graduated as an engineer in 1822. He worked for a short time as mine manager in Alsace. When the state of Colombia was created, Boussingault left for South America and Bogotá where the new government appointed him professor in a school for the training of engineers. He arrived in Bogotá in the midst of an insurrection and the government appointed him colonel in the army. This appointment did not prevent him from pursuing a number of research projects in different fields: mineralogy, geology, plant chemistry, climatology. The results of this activity he sent to von Humboldt who presented them at the Académie des Sciences of Paris. Therefore when he returned to France in 1832, Boussingault had won for himself a reputation as a scientist.

Though deprived of any academic qualification (he did not even hold a bachelor's degree), Boussingault was allowed to present a doctorate thesis on the process of silver isolation by amalgamation, as practised in America. He was appointed professor of chemistry at the Faculty of Sciences of Lyon, and the following year became dean of this Faculty where he remained until 1837. He was at that time called to Paris as "suppléant" of Thenard at the Faculty of Sciences and, in 1839, he became a member of the Académie des Sciences. His reputation as an agricultural chemist was such that a special chair of "Economie rurale" at the Conservatoire des Arts et Métiers was created for him, which he held until his death in 1887.

During the decade he spent in South America, Boussingault had been struck by the extraordinary fertilizing influence of Peruvian guano, composed mainly of ammonium salts. When he became, by marriage, the owner of a farm in Alsace, at Bechelbronn (now Pechelbron), * he was able to organize vast experiments in the

---

* He married a daughter of a rich landowner, Le Bel, whose family exploited oil wells (now exhausted) at Pechelbronn. Boussingault's wife was the aunt of J.A. Le Bel, who gained fame for his work in stereochemistry.

field (literature in Aulie [52]). Boussingault was uncertain about the reigning theories of the origin of plant nitrogen. In 1837 and 1838 he carried out trials, seeding plants in a material which was calcined, with the objective of destroying any nitrogenous matter (but also destroying bacteria, unknown to him at that time). He observed that clover and peas extracted nitrogen from the air, but that wheat and oats did not. On his farm (1838—1841) he demonstrated by crop rotation, that legumes restore the nitrogen of the soil. He therefore was able to identify the well-known empirical effect of legumes, for instance of clover, with an effect on the nitrogen content of the soil. But Boussingault left open the mechanism of this action, which he considered as possibly due to the ammonia of the air. He denied, and continued to deny until the end of his life, that the nitrogen could result from an absorption of atmospheric nitrogen through the leaves.

18. Return to a search for atmospheric ammonia (1854—1856) and demise of the ammonia theory

As stated above, ammonia had, from the start, been considered by several authors, including Boussingault, as a possible source of nitrogen from the air.

In 1847, Péligot [60] published a method for measuring combined nitrogen in solution by means of titration against a known acid. Boussingault relied on the method to fulfill a wish he had expressed in 1853. Following a balance sheet method, he demonstrated that the nitrogen brought down in the form of ammonia from the atmosphere is adequate to supply but a small part of the nitrogen needs of plants (literature in Aulie [52]). This conclusion, which dismissed the ammonia theory, was confirmed in England by Way [61], who wrote:

> "The refreshing influence on vegetation of a thunderstorm is due to the much needed water *as* water, and not to its being a vehicle of nitrogenous manure" (p. 621).

19. If plant nitrogen does not come from atmospheric ammonia, where does it come from? *

That plants assimilate the free gaseous nitrogen of the air had been claimed by Ville in 1852 [62]. A former student of Boussingault at the Conservatoire des Arts et Métiers, Ville [58] became demonstrator to Boussingault there in 1847. A first dispute arose between Boussingault and Ville concerning the accusation raised by Boussingault against his demonstrator, of overspending the funds of the laboratory and using them for his private research, an accusation which may have been exaggerated by personal dislike, as Boussingault spent half the year in Alsace. However, Boussingault never doubted the competence of Ville as an analyst. Ville resigned. He obtained a modest teaching post at the Couvent des Carmes in the garden of which he started research on the possible assimilation of gaseous nitrogen by plants. How, from this obscure post, he became professor of "Physique végétale" at the Museum of Natural History deserves a short commentary, explaining also why Ville was so heartily disliked in the academic community. Ville (1824—1897) [58] was an illegitimate son of King Louis of Holland, brother of Napoleon, and of a housemaid of Queen Hortense. His mother having been married to a member of the imperial police, Georges Ville of Lyon, the child acquired his name. When after the coup d'état of 1851 Napoleon III became emperor, he raised his step-brother, in 1852, from his post at the Couvent des Carmes to a chair at the Institut d'Agronomie Végétale of Versailles. When Napoleon III confiscated the land of this Institute for his hunting parties, in 1857 he created for his step-brother a new chair (of "Physique végétale") at the Museum of Natural History, without any approval from the council of professors. The professorial staff of the Museum threatened to resign but were made aware that their resignation would be accepted and therefore abandoned the action (McCosh [58]).

Imperial favor was showered on Ville in other ways. He received a large piece of land in Vincennes in order to perform his experiments on fertilizers. He also owned land in the northern district of

* In this section and the remaining part of this Chapter, the text is based on studies concerning the primary literature of the subject (references and analysis) by Aulie [52] and by McCosh [58].

References p. 62

Plate 132. Georges Ville.

Paris where, near the Arc de Triomphe, the rue Georges-Ville, a name given by him, now connects avenue Victor-Hugo and rue Paul-Valéry. As a consequence of his experiments (1849–1850) in the garden of the Couvent des Carmes, Ville had claimed that he had demonstrated a direct assimilation of the nitrogen of the air by plants. In these experiments Ville [63] grew plants in a bell-jar sealed at the base with water. Unwashed air charged with carbon dioxide was continuously renewed. In 1849, in such experiments, he grew cress and lupins and obtained an increase in nitrogen content of lupin but not of cress. In 1950, in similar experiments, he succeeded in observing an increase in nitrogen content in plants of sunflower, tobacco, colza and wheat. In these experiments, he flushed his apparatus with air washed with sulfuric acid to remove ammonia.

Boussingault's second study of nitrogen fixation took place between 1851 and 1854. Since his first experiments in 1838, chemical advances in soil chemistry had been made. It was known that nitrates were formed in the soil by decomposition and that ammonia was a step in the process.

In his second set of experiments, Boussingault [64] took great care to avoid any trace of nitrogenous matter in the soil. From the theoretical point of view, if nitrogen was directly absorbed by the leaves as Ville believed, there was no reason to add nitrogenous compounds as fertilizers, the value of which, as Liebig claimed, resided only in their mineral salts. On the other hand, if there was no fixation, soil must be rendered fertile by an addition of both mineral salts and nitrogenous compounds, as Boussingault claimed on the basis of his farming experiments. Therefore Boussingault designed an apparatus in which he used for soil a crushed, calcined pumice. He mixed it with the ash of unspecified manure or the ash of the plant species tested, and he ground the seeds in that medium in porcelain crucibles. By the technique of Péligot he determined the amount of nitrogen in the entire plant, in the whole soil with its root fragments, and even in the crucible. He had taken particular care to avoid any possibility of contamination by atmospheric ammonia. He concluded that the nitrogen gas of the air had not been assimilated by beans, cress or lupin.

At the meeting of the Académie des Sciences of April 10, 1854 Ville [65] maintained his own conclusions. The reason for Boussingault's failure to detect nitrogen fixation was due, Ville claimed,

References p. 62

to the nature of his methodology. He used non-renewable atmosphere. Ville also criticized the use of porcelain crucibles which, according to him, prevented a normal life of the plants. Curiously enough, during the heated discussion at the Academy, Ville mentioned that Boussingault, who was now denying nitrogen fixation by plants, had discovered it in 1853. In fact, as we have stated, Boussingault had demonstrated the increase in soil nitrogen after legume cultivation, but he had denied the absorption of atmospheric nitrogen by plant leaves.

Boussingault [66] replied:

"I do not have to defend myself against an attack on my work: far from that, I have to defend myself from having made a discovery" (translation by McCosh [58]).

In the same discussion, as pointed out by McCosh [58], Boussingault declared that he had always observed a gain in nitrogen when the soil contained the most trifling quantity of nitrogenous matter, but no gain if no nitrogen was left in the soil. This points to the interest which was raised with respect to what we now call nitrification in the soil.

As the conclusions of Ville and of Boussingault were in complete opposition, Ville offered to repeat his experiments before a committee appointed by the Academy, a process which was commonly relied upon at the time in France. The committee appointed under the chairmanship of Chevreul consisted of Dumas, Regnault, Payen, J. Decaisne (a horticulturist professor at the Museum of Natural History) and Péligot (McCosh [58]), assisted by a chemist, S. Cloëz (assistant of Chevreul at the Museum of Natural History). This committee, though not endorsing the conclusions of the assimilation of gaseous atmospheric nitrogen by plants, concurred that the experiments performed by the commission agreed with Ville's results (see McCosh [58]).

Boussingault was shocked by the fact that Ville's experiments had been chosen for verification. (Letter of Boussingault to Dumas, June 23, 1854. Académie des Sciences, dossier Boussingault, read by McCosh [58], note 54.) He resented this to the extent of considering resigning his position in Paris for the rectorship of the Faculty of Strasbourg. He considered it as a lack of goodwill towards him. (Boussingault to Péligot, 1801—1854. Dossier Boussingault. Quoted after McCosh [58].)

In the course of the discussion at the Academy in April 1854

Boussingault had announced new experimental facts. These came out in a third study, the results of which he presented at the Academy on October 2, 1854 [67] and published in an extensive paper of the same year [68]. He answered the criticism of Ville and maintained his claim that plants do not, in the conditions of his experiments (analyzed in detail by Aulie) [52], absorb gaseous nitrogen even in a constantly renewed atmosphere.

Ville, in a book [69] published in 1857, was not as positive as he had been in previous writings, with respect to nitrogen fixation by plants. He could no longer observe the striking gains of nitrogen he had recorded before.

In several countries, the query which had fed the controversy between Ville and Boussingault interested a number of workers. One of the best organized researches was that of Lawes, Gilbert and Pugh [70] published in 1861. In those experiments a refinement was introduced in comparison with the technical methodology used by Boussingault and by Ville: the washed air was kept at a pressure slightly above atmospheric pressure. This prevented the entry of air from outside the apparatus. These well-controlled experiments accomplished at Rothamsted gave no evidence of nitrogen fixation by plants. The controversy between Boussingault and Ville is, as McCosh [58] writes:

"an example of the failure of an accurately controlled experiment which omits an important factor".

This factor was the presence, in natural conditions, of bacteria, lacking in the experiments. If Boussingault obtained negative results, it was due to his careful elimination of contaminants such as ammonia or dust and to his insistence on using calcined pumice as soil, a precaution which also made the soil bacteriologically sterile. This is why he ignored one of the main components of the nitrogen cycle: the fixation of gaseous nitrogen by a category of soil bacteria, unknown at the time.

On the other hand, the results obtained by Ville may have been due to a massive introduction in the ground of his pots, through the aspiration of large volumes of air at great speed, of nitrogen-fixing bacteria. The concept of nitrogen-fixing bacteria was, as we shall see, introduced only thirty years later. In the meantime, mainly owing to the work of Kuhlmann, at Lille, the attention was shifted, from the air and plant foliage, to the soil and roots.

References p. 62

Plate 133. Frédéric Kuhlmann.

20. Soil as a chemical system

It had been stated by Liebig that nitrates introduced in soil had a beneficial effect, but Liebig denied that this effect was due to the nitrogen.

Kuhlmann, professor of industrial chemistry at the University of Lille, had published in 1846 [71] and 1847 [72] results of the application of ammonium compounds (ammonium chloride, ammonium sulfate, ammonium nitrate, horse urine, ammoniacal liquor from gas) on the grass of a meadow producing hay and obtained remarkable increases in plant growth. As he always associated this with the presence of putrescible material in the soil, he evolved the theory known as "putrid fermentation", according to which in the soil, nitrates had first to be converted to ammonium compounds, before being assimilated by plants. Of course, Kuhlmann was unaware of the role of bacteria and he attempted to provide a purely chemical explanation which is epitomized by Aulie [52] as follows:

"Noting the importance of alkalinity and solubility, the presence of ammonium carbonates in the soil, the action of sulfuric acid on saltpeter, and the role of porosity as a catalytic *milieu*, he suggested that ammonium salts might be formed in the soil by a deoxidation of nitric acid, in a reaction similar to the following (Kuhlmann [71], p. 228):

$9\ ClH + 8\ Zn + NO_5CuO = 8\ ClZn + ClH, \quad NH_3 + Cu + 6\ H_2O$"

(p. 464, footnote 143).

He also recognized the importance of nitrification in the superficial layers of the soil as a kind of inverse reaction, that is primarily oxidative, and he suggested the following schema (Kuhlmann [71], p. 233):

$$8\ MnO_2 + SO_3NH_3HO + 7\ SO_3HO = 8\ SO_3MnO + NO_5HO + 10\ HO$$

In 1856, Boussingault [73] showed that the process of "putrid fermentation" was unnecessary. This he showed by growing sunflowers in pots containing sterile soil. When saltpeter was added, the plants were developing well in contrast to the poor growth of those not provided with the saltpeter. Turning to the effect of several salts, he added them to sterile soil and demonstrated without any doubt the role played by nitrates in plant nutrition [74, 75]. But it remained unknown at the time in which form nitrogen

References p. 62

ДП

from the soil really entered the plant.

Ville [76] had also pursued research related to the effects of mineral salts on plant growth. He used artificial soils made up of mixtures of inorganic compounds. Ville concluded that, though mineral salts alone or nitrates alone produced little effect, the greatest effect was observed when a phosphate was employed in conjunction with a nitrate [69]. Boussingault, at the meeting of November 23, 1857, claimed priority on this point and a controversy was raised between them. From the evidence available we may conclude as McCosh [58] does:

"It is enough to say, that Ville and Boussingault independently demonstrated the combined effect of nitrates and phosphates on plant growth" (p. 489).

Aulie [52] has analyzed in detail the work of Boussingault in the field of nitrification.

21. Nitrification as a bacterial process

In 1877, J.J.T. Schloesing and Müntz [17], building on a suggestion of Pasteur [78], pointed to the role of the "ferment nitrique" in the process. The background of the work was the purification of sewer water. The traditional theory, derived from Liebig, was that nitrification was a combination of oxygen with the nitrogenous material, while, as suggested by Pasteur, nitrification could result from an action of the "ferment nitrique". Schloesing and Müntz allowed sewer water to seep slowly through a column of sand in the presence of flowing oxygen and they noted the disappearance of ammonia. They showed that this process is interrupted by chloroform and reactivated by adding a pinch of "terre végétale". They concluded (though they did not succeed in isolating the bacteria involved) that only the "ferment nitrique" could be considered.

22. The discovery of nitrogen-fixing bacteria in the soil

This development was foreseen by Berthelot [79] who published in 1882 experiments showing that samples of soils spontaneously

References p. 62

get richer in nitrogen, while controls with soil heated to more than 100°C showed no such increase. He suggested the participation of nitrogen-fixing bacteria.

It was in 1893 that Winogradsky [80] reported the existence in soils of the nitrogen-fixing bacterium *Clostridium pasteurianum*. Other nitrogen-fixing microorganisms were discovered, e.g. *Azotobacter*.

These discoveries completed the concept of the nitrogen-fixing bacteria of the soil in which the nitrogen is fixed as ammonium compounds converted by other bacteria to nitrites and finally to nitrates acting as plant nutrients.

23. The special case of legumes

When nitrogen gas is circulated through a closed system over legumes growing in sterile soil, a diminution of nitrogen is observed (A.T. Schloesing and Laurent [82]). That in the case of legumes the fixation of nitrogen is accomplished by symbiotic bacteria contained in nodules on roots of legumes, was documented later, as we shall see.

24. Retrospect

In this chapter, we have retraced the origin of the concept of the aerial nutrition of green plants, directly from the carbon dioxide of the air (exclusive source of the carbon) and indirectly from the nitrogen of the air (partial but important source of the nitrogen of plants).

When the aerial origin of plant carbon was recognized, the concept became current that the atmosphere constitutes a link between the vegetable and animal kingdoms. Plants obtaining from air the precursors of their constituents, a feature considered as lacking in animals, plant constituents were considered as the natural food of herbivorous animals. On the other hand the supply of "respiratory food" to animals was considered as the source of the *pabulum* of plants, liberated by animals as metabolic products.

Before dealing with this scheme limiting to plants the faculty of biosynthesis, we shall consider the archeology of the long lasting concept of the relation between vegetable food and animal flesh, considered as deprived of biosynthetic powers.

References p. 62

REFERENCES

1 J.F. Fulton, Isis, 18 (1932) 77.
2 S. Hales, Vegetable Staticks, London 1727 (the 6th chapter is entitled Analysis of Air).
3 J.E. Harvard and W.C. Thomas, Am. J. Med., 45 (1968) 693.
4 Lectures on the Elements of Chemistry delivered in the University of Edinburgh by the late Joseph Black. Now published from his manuscript by John Robison, 2 vol., Edinburgh, 1803.
5 H. Guerlac, Isis, 48 (1957) 124, 453.
6 H. Guerlac in C.C. Gillispie (Ed.), Dict. of Sci. Biogr., Vol. 2, New York, 1970, p. 173.
7 S. Hales, Statical Essays, 2 vol., London, 1733.
8 N.G. Coley, Ambix, 18 (1971) 69.
9 S. Hales, An Account of Some Experiments and Observations on Mrs. Stephen's Medicines for Dissolving the Stone, London, n.d. (1741 according to Guerlac [5], p. 139, footnote 65).
10 J. Black, Experiments upon Magnesia Alba, Quicklime and some other alkaline substances. Essays and Observations. . ., 2 (1756) 157.
11 D. MacBride, Experimental Essays on Medical and Philosophical Subjects . . ., Dublin, 1764.
12 H. Cavendish, Phil. Trans. Roy. Soc., 56 (1767) 141.
13 J.B. Priestley, Phil. Trans. Roy. Soc., 62 (1772) 147.
14 R.E. Schofield, in C.C. Gillispie (Ed.), Dict. of Sci. Biogr., vol. 11, New York, 1975, p. 139.
15 F.W. Gibbs, Joseph Priestley, Adventurer in Science and Champion of Truth, London, 1965.
16 L.K. Nash, in J.B. Conant and L.K. Nash (Eds.), Harvard Case Histories in Experimental Science, Vol. 2, Cambridge, Mass., 1957, p. 323.
17 D.W. Singer, Ann. Sci., 6 (1949) 127.
18 T. Percival, Medical Ethics, Manchester, 1803.
19 F.H. Garrison and L.T. Morton, A Medical Bibliography, 2nd ed., revised, London, 1965.
20 Essays Physical and Chemical by M. Lavoisier, London 1776.
21 T. Percival, Philosophical, Medical and Experimental Essays, London, 1776.
22 N.T. de Saussure, Recherches Chimiques sur la Végétation, Paris, 1804.
23 E.L. Scott, Ambix, 17 (1970) 43.
24 J.B. Priestley, Experiments and Observations Relating to Various Branches of Natural Philosophy, Vol. 1, London, 1779.
25 J.B. Priestley, Experiments and Observations Relating to Various Branches of Natural Philosophy, Vol. 2, London, 1781.
26 J. Ingen-Housz, Experiments upon Vegetables Discovering their Great Power of Purifying the Common Air in Sunshine and Injuring It in the Shade and at Night, London, 1779 (Reprint in Chronica Botanica, 11 (1948) 287.) A French enlarged version appeared in the following year: Expériences sur les Végétaux, Paris 1780. A revised and enlarged edition

appeared later: Expériences sur les Végétaux, Vol. 1, Paris 1787.
27 P.W. Van der Pas in C.C. Gillispie (Ed.), Dict. of Sci. Biogr., Vol. 7, New York, 1973, p. 11.
28 J. Ingen-Housz, Phil. Trans. Roy. Soc., 70 (1780) 154.
29 Ch. Bonnet, Recherches sur les Fonctions des Feuilles, Paris, 1754.
30 S. Ruben, M. Randall, M. Kamen and S.L. Hyde, J. Am. Chem. Soc., 63 (1941) 877.
31 J. Sénebier, Mémoires Physico-chimiques sur l'Influence de la Lumière Solaire pour Modifier les Etres des Trois Règnes de la Nature et surtout ceux du Règne Végétal, 3 vol., Geneva, 1782.
32 J. Sénebier, Recherches sur l'Influence de la Lumière Solaire pour Métamorphoser l'Air Fixe en Air Pur par Végétation, Geneva, 1783.
33 P.E. Pilet, Arch. Int. Hist. Sci., 15 (1962) 303.
34 P.E. Pilet, in C.C. Gillispie (Ed.), Dict. of Sci. Biogr., Vol. XII, New York, 1975, p. 308.
35 J. Briquet, Bull. Soc. Bot. Suisse, 53a (1940) 433.
36 J. Sénebier, Expériences sur l'Action de la lumière Solaire dans la Végétation, Geneva, 1788.
37 J. Sénebier, Observations sur la Physique, 41 (1792) 205.
38 J. Ingen-Housz, Expériences sur les Végétaux, Vol. 2, Paris, 1789.
39 J. Ingen-Housz, Ernährung der Pflanzen und Fruchtbarkeit des Bodens, Leipzig, 1798.
40 E.A. Letts and R.F. Black, Sci. Proc. Roy. Dublin Soc., 9(II) (1900) 105.
41 E.I. Rabinowitch, Photosynthesis and Related Processes, Vol. I, New York, 1945.
42 J.H. Hassenfratz, Observations sur la Physique, 13 (1792) 178, 318; 14 (1792) 55.
43 J. Liebig, Chemistry in its Application to Agriculture and Physiology, London, 1840.
44 H. von Mohl, Grundzüge der Anatomie und Physiologie der vegetabilischen Zellen, Braunschweig, 1851.
45 J. Sachs, Lehrbuch der Botanik, Leipzig, 1868.
46 J. Moll, Landwirtsch. Jahrb., 6 (1877) 327.
47 J. Moll. Arb. Bot. Inst. Wurzburg, 2 (1878) 105.
48 F.J.A.N. Unger, Anatomie und Physiologie der Pflanzen, Pest, Vienna and Leipzig, 1855.
49 J. Sachs, Bot. Ztg., 20 (1862) 365; 22 (1864) 289.
50 H. von Mohl, Vermischte Schriften botanischen Inhalts, Tübingen, 1845.
51 C. von Nägeli, Die Stärkekörner, Zürich, 1858.
52 R.P. Aulie, Proc. Am. Phil. Soc., 114 (1970) 435.
53 C.L. Berthollet, Mém. Acad. Roy. Sci., (1780) 120 (published 1744).
54 C.L. Berthollet, Mém. Acad. Roy. des Sci. (1785) 316 (published 1788).
55 A.F. de Fourcroy, Ann. Chim. Phys., 3 (1789) 252.
56 J.A. Chaptal, Chimie Appliquée à l'Agriculture, Paris, 1829. (English translation: Chemistry Applied to Agriculture, New York, 1840).
57 H. Gautier, Boussingault (Jean-Baptiste-Joseph-Dieudonné) 1802—1887 in L'air, l'Acide Carbonique et l'eau. Mémoires de Dumas, Stas, Boussingault (Les Classiques de la Science, No. 1), Paris, 1913.

58 F.W.J. McCosh, Ann. Sci., 32 (1975) 475.
59 R.P. Aulie, in C.C. Gillispie (Ed.), Dict. of Sci. Biogr., Vol. 2, New York, 1970, p. 356.
60 E.M. Péligot, J. Pharm. Chim., 11 (1847) 334.
61 T.T. Way, J. Roy. Agric. Soc. England, 17 (1856) 129, 618.
62 G. Ville, Compt. Rend., 35 (1852) 464.
63 G. Ville, Compt. Rend., 31 (1850) 578.
64 J.B. Boussingault, Ann. Sci. Nat. (Bot.), ser. 4, 1 (1854) 241.
65 G. Ville, Compt. Rend., 38 (1854) 705, 723.
66 J.B. Boussingault, Compt. Rend., 38 (1854) 717.
67 J.B. Boussingault, Compt. Rend., 39 (1854) 501.
68 J.B. Boussingault, Ann. Sci. Nat. (Bot.), ser. 4, 2 (1854) 357.
69 G. Ville, Recherches Expérimentales sur la Végétation, Paris, 1857.
70 J.B. Lawes, J. Gilbert and E. Pugh, Phil. Trans. Roy. Soc., 151 (1861) 431.
71 C.F. Kuhlmann, Ann. Chim. Phys., (3) 18 (1846) 138.
72 C.F. Kuhlmann, Ann. Chim. Phys., (3) 20 (1847) 223.
73 J.B. Boussingault, Ann. Chim. Phys., (3) 46 (1856) 5.
74 J.B. Boussingault, Compt. Rend., 44 (1857) 940.
75 J.B. Boussingault, Compt. Rend., 47 (1858) 807.
76 G. Ville, Compt. Rend., 45 (1857) 996.
77 J.J.T. Schloesing and A. Müntz, Compt. Rend., 84 (1877) 301.
78 L. Pasteur, Compt. Rend., 44 (1862) 265.
79 M. Berthelot, Compt. Rend., 115 (1882) 738.
80 S. Winogradsky, Microbiologie du Sol, Paris, 1949.

Chapter 41

From Vegetable Food to Animal Flesh

1. Globulist (globularist) theories of tissue structure and tissue formation

The biosynthetic pathways which have been individualized today were at first hidden behind the material reality of much broader complexities. In the case of animals, the concept of biogenesis did at the start concern the whole individual and it was progressively that smaller parts became the subject of research; the pathway, instead of extending from the matter of food to the whole organism, going from food to tissue, from food to cells. Later the path was located between the "proximate principles" of the food and those of the cells. That the reality covered by the term"nutrition" has changed with time has been noted by Holmes [1].

The process of reduction involved in these changes has consisted in a process of focussing from a broader complex material to a smaller one, finally reaching an assembly line effected by enzymes after having started from the concept of a direct intercalation of food material into animal flesh.

At the beginning of the 19th century, a number of microscopists believed that they could recognize, under the microscope, visible particles in animal tissues. These particles they called "globules", sometimes "molecules". After a series of disconnected observations, a theory of animal texture based on "globules" was first formulated by Wolff in 1759 and incorporated by Meckel in his *Textbook of Anatomy* (literature in Baker [2]). The "globules" were generally considered as embedded in a homogeneous matrix.

The "globulist" theory, a form of the theory of "direct assimilation" (see Part I, *passim)*, was an aspect of mechanical physiology, a particle of food replacing a worn out particle of flesh. Dur-

Plate 135. John Hunter.

ing the early 1820's, in France, under the development of Bichat's work, "general anatomy" was based on this mechanical concept, in which a mechanical view of "development" took precedence over chemistry. The methodology was essentially based upon the use of the microscope and mainly on the study of embryonic tissues.

2. Blood coagulation considered as necessary to tissue formation

In England, clinical preoccupations were at the origin of superimposing a vague chemical flavor on the microscopical observations, generally based on optical illusions. This development was influenced by the recurring and overgeneralized concept of "coagulation", recognized by Bachelard [3] as one of the worst themes of generality, in which the most heteroclitous range of natural facts has been abusively collected. Coagulation has been considered as the explanation of the formation of animals by the famous geologist Wallerius [4] who quotes from Job:

"Instar lactis me mulxisti, et instar casei coagulari permisisti".

It was a practical surgical problem, the healing of wounds, which was the occasion for superimposing the vaguely chemical coagulation concept on the mechanical globulist concept. John Hunter, in a book on blood, inflammation and gun-shot wounds, published in 1817, considered that blood is formed in all tissues of the animal body. He believed that the coagulation of blood is a necessary prelude to tissue formation. Rather [6] has pointed to the relation of this concept with the revival of humoralism, which marked the period around 1800. In an article where the reader will find a bibliography of the forms taken by the globulist theory in different national contexts, Pickstone [7] has commented on the genetic epistemology of Hunter's theory, as follows:

"The whole of Hunter's approach depended on his readiness (common to those who later adhered to the blasteme theories) to accord life to apparently unorganized fluids. Though this idea had originally (1755 and 1756) sprung from his consideration of the self-preserving power of the parts of a hen's egg and though Hunter and his students were deeply interested in generation, the problem of tissue formation, as formulated by them, continued to centre on the relationship between blood and the mature tissues rather than on embryological development" (p. 340).

Hunter and his student Hewson considered blood clotting to be

References p. 85

Plate 136. Sir Everard Home.

due to a coagulation of the "fibrin" (not our fibrin) in the plasma and not to an aggregation of red blood corpuscles. This approach to wound healing on the basis of the formation of a fibrin clot proved to be a fruitful concept in the development of the knowledge of cicatrization, but from this sound basis, a student of John Hunter, the surgeon Sir Everard Home [8], was led astray by microscopical illusions.

To quote Baker [2] (p. 119):

"His studies were made in collaboration with a Mr. Bauer who seems to have done most of the practical work. They noticed (1818) that when blood coagulates, the red blood corpuscles tend to unite in line. They then boiled or roasted voluntary and involuntary muscle, macerated it in water, and found that the "fibers" are readily broken down into a mass of globules of the size of those in the blood, deprived of their color" (p. 175).

Mazumdar [9] epitomizes the conclusions of Home as follows:

"If, as Home supposed, the globules consisted of fibrin, the loss of the red material seemed to take place when the clot formed would allow the fibrin globules to stick together and form first a fibrin strand, then a muscle fiber. The decrease in size of the corpuscles which had lost their red matter brought them into the size range of the globules of chyle, those of the coagulable lymph, and those of the muscle fiber. . . The system was complete, from food to flesh" (p. 246).

The theory was rejoining with the basic concepts of *Naturphilosophie.* Oken [10] had for instance maintained that the formation of the body material of animals took place in blood and lymph. Before, in 1796, Reil had stated that the red blood corpuscles.

"lined up to form fibrin strands from which animal tissues and the organs are composed" (cited after Mazumdar [9]).

Pickstone [7] has wisely noted that if chemistry added little to the contents of Home's theory, from the point of view of genetic epistemology,

"it was contemporary chemistry that had focussed attention on the relationship between the blood and the tissues" (pp. 342—343).

The "analysts", as was stated in Chapter 5, had been concerned with the changes in elementary composition of blood, lymph and tissues and pointed to their relationships as well as with food (see also Holmes [11]).

Pickstone [7] (footnote, p. 343) notes that Fourcroy [12], when comparing the elementary composition of blood and tissues,

Plate 137. Jean Louis Prévost.

quotes the Hippocratic notion, as expressed by T. de Bordeu, that the blood was "flowing or fluid flesh".

When the proximate (immediate) principles were individualized, in spite of their imprecise chemical definition, the chemists endeavored to recognize where they first appeared. They tried to locate this emergence along the classical sequence: food → chyme → chyle → blood → tissues (see Part I, *passim*).

As stated above, it had been claimed by John Hunter and by Hewson that "fibrin" is dissolved in the fluid phase of blood, while Home considered that it was contained in the red corpuscles. Prévost and Dumas situated it more specifically in the nucleus (in frog's blood with nucleated red blood cells). Prévost [13] and his collaborator Dumas, who was later to become a leader in organic chemistry, had been pursuing studies on fertilization and development in vertebrates. Prévost was practising medicine in Geneva and he devoted his spare time to experimental studies. Dumas, born in 1800 in Alais (now Alès) in the south of France, had left his native town to go to Geneva where he intended to learn the art of pharmacist from Leroyer, a friend of Prévost, who associated Dumas with his own studies (see Brunner [13]).

One of the papers of Prévost and Dumas [14] dealt with the action of blood in different aspects of life. The authors stated that it was the nucleus of the red cells of frogs which went to make the fibers of the blood clot. Submitting egg-white to a galvanic current, the authors reported a tendency to form rows of "globules". This paper was highly praised in scientific circles. One of the results reported certainly deserved this praise: the discovery, in blood, of the presence of urea which was known to occur in urine.

It was after reading this paper that Humboldt paid a visit to the young Dumas in Geneva. He advised him to move to Paris where his talents would find more adequate surroundings. Warmly recommended to Arago, Dumas was, at the age of 24 years, appointed as "préparateur" of Thenard at the Chemical Laboratory of the Ecole Polytechnique. That the genesis of tissues consisted in the aggregation of entities visible under the microscope was also proposed by Edwards [15,16] *. He claimed that all ani-

* Henri-Milne Edwards, brother of the physiologist F.W. Edwards (mentioned in Chapter 8, p. 165) is often, in bibliographies erroneously called H. Milne-Edwards, a patronymic name which was taken after his death by his son Alphonse (see this Treatise, vol. 29B, p. 234, note).

References p. 85

Plate 138. Henri-Milne Edwards.

mal tissues were composed of spherical globules of 1/300 mm in diameter. He also stated that nutrition was a distribution of red blood cells becoming inserted between already present globules.

Berzelius, in his treatise on animal chemistry, a German version of which was published in 1831 [17], although recognizing that lymph, though deprived of corpuscles, would clot, still referred to hemolyzed red blood cells, deprived of their pigment, as "fibrin corpuscles". This is another reason for avoiding, in the literature of the time, any identification of "fibrin" with a well-defined "proximate principle" such as our fibrinogen.

In the thirties, the common opinion reigned according to which the tissues were composed of "globules".

As it was known that there are corpuscles in blood and in chyle, it was considered that the chyle corpuscles turn into blood corpuscles as they enter the circulation, and finally become flesh. In their paper of 1821, in harmony with this theory, Prévost and Dumas [14] reported an alleged production of rows of corpuscles in egg-white submitted to a galvanic current, and called the attention, in relation to the formation of the chyle corpuscle, to this remarkable phenomenon which indeed, for a short time, was considered as a milestone.

3. Dismissal of globulism

As stated above, while the presence in solution of fibrin in the fluid phase of blood had been accepted by Hunter and by Hewson, the common opinion, based on the writings of Home, of Prévost and Dumas and of Edwards [15,16] was that "nutrition" consisted in the aggregation of particles visible under the microscope. It is this view which is contradicted by the first sentence of the chapter on Nutrition in the *Handbuch* [18] of Müller:

"Die Ernährung ist kein Gegenstand mikroskopischer Beobachtung".

By writing that

"Nutrition is not an object of microscopical research",

Müller dismissed the "intercalatory" * theory of the globulists

* This term, which (as mentioned in the *Oxford Unabridged Dictionary*) had been used for the first time in a translation of Augustine, was, in the present context, used by Rather [6] in his book on Addison. Rather notes that the theory was again proposed in 1840 when Addison considered that white corpuscles take part in tissue formation.

References p. 85

Plate 139. Johannes Müller.

and introduced the chemical concept of a passage of dissolved nutrients into the blood through capillary walls. In the *Handbuch* Müller introduces a clear distinction between plasma (*liquor seu lympha sanguinis*) and the serum exuded by the clot (*coagulum sanguinis*). Serum contains serum albumin, heat clotted at 70°C. When coagulation takes place, the plasma divides up between fibrin, previously in a dissolved state, and serum.

To quote from the *Handbuch* [18]:

"The usual opinion about the clotting of the blood is that the red clot is formed by the aggregation of red corpuscles and that the nuclei of the red corpuscles are themselves the fibrin (a concept introduced by Prévost and Dumas), which was clothed in a hull of coloured material, which after clotting was washed off, leaving a white coagulum... Berzelius has suggested that the lymph holds dissolved fibrin, and so the blood must too, as they are similar fluids. One can find an even better reason, that the lymph itself goes into the blood. The idea that fibrin is dissolved in the blood, had already been suggested from time to time. I have been so lucky as to find a definite proof for Berzelius's view, and I am in a position to show that the red clot is a mixture of fibrin that was dissolved before, and of red corpuscles" (p. 105) (translated by Mazumdar [9], p. 243).

Müller refers to a number of experiments by which he demonstrated that "fibrin" is dissolved in the *liquor sanguinis*. For instance, he succeeded in getting a filtrate of plasma of frog's blood and observed that it did clot, though deprived of corpuscles. These filtration experiments were first published in 1832 [19]. Reporting on these experiments, Berzelius [20] pointed to the fact that they disproved the views of Home, as well as those of Prévost and Dumas (see Mazumdar [9]).

Müller had measured the diameter of the blood corpuscles and he had shown that they were larger than the globules supposed to be the components of muscle. But Home had taken care of this in advance by claiming that a reduction of size took place in the process.

The crucial experiment on blood coagulation showing that dissolved fibrin in the plasma, rather than globules, constitutes the coagulable part of the blood, led Müller to reject the globulist approach of "nutrition".

But the decisive argument against the mechanical theory of "nutrition" was the discovery by Schwann of the membrane lining capillary walls (reported in Müller's *Handbuch* [18]). As there are no gaps in capillary walls, the nutrition material, in order to reach

References p. 85

Plate 140. Mathias Jakob Schleiden.

the blood, must cross the membrane in a state of solution. This material can therefore only be identified by chemical methods.

The concept according to which the "proximate principles" reached the blood by crossing membrane frontiers in a state of dissolution dismissed the direct assimilation from food but it did not exclude "intercalatory" theories altogether. For instance, Prévost and Dumas [14], as noted above, considered the possibility of the formation in chyle, from dissolved constituents, of corpuscles later transformed into red blood cells and intercalated.

But the "intercalarist" concept of an introduction between the "globules" of tissues, of globules of "albumen" or other plant or animal bodies, was finally undermined and discarded as a consequence of perfecting the quality of microscopes. Using such an instrument, Hodgkin and Listar [21] (1827) looked in vain for "globules" in different tissues. This experience was shared by a number of microscopists, including Schwann. In his unpublished laboratory notebook, Schwann [22] records, on April 2, 1835, that he had found in direct observation no grounds to believe that the primitive fibers of muscle are composed of globules.

4. Schleiden's "cell theory"

We have, in Chapter 6, retraced the historical importance, in the establishment of the cell theory, of the theory of fibres, and *later* of the views of the globulists. The concept that plants were composed only of nucleated cells, of modified cells and of products of cells became current in the 1830's. Though many authors, besides Schleiden, have recognized the cellular nature of plant texture, he is often considered erroneously as the introducer of this notion. The importance of Schleiden's contribution is mainly linked with his opposition to "Naturphilosophie" as clearly expressed in his *Grundzüge der Wissenschaftlichen Botanik* [23]. But his fame mainly rests on his formulation of what became known as the cellular theory, which lasted only a short time, between its formulation in 1838 with respect to plants, and 1877, the date of the formulation by Virchow, of the concept expressed by the phrase "*Omnis cellula e cellula*". This episode started with the publication by Schleiden [24] (1838) of a paper on phytogenesis. At the time, it was already known, as was stated in Chapter 6, that plants

References p. 85

were composed of nucleated cells. The nucleus which had been discovered in plant cells by Brown, was called "cytoblast" by Schleiden, "in reference to its functions", i.e. its assumed role in cell formation. Schleiden describes what we now call nucleolus and states that it is formed earlier than the cytoblast.

To quote Schleiden:

"So soon as the cytoblasts have attained their full size, a delicate transparent vesicle rises upon their surface. This is the young cell, which at first represents a very flat segment of a sphere, the plane side of which is formed by the cytoblast, and the convex side by the young cell, which is placed upon it somewhat like a watch-glass upon a watch" (see p. 136 of Vol. 30, footnote).

The main argument of Scheiden is that he "found cell nuclei previous to the appearance of the cells, floating loose in the fluids in very many plants" (p. 210). This is a correct observation which has been more recently duplicated, for instance, by the recognition of loose nuclei in coconut milk.

Schleiden does clearly state that the cell theory is not applicable to animals and he explains why:

"1. The plant grows, that is it produces the number of cells allotted to it.

2. The plant unfolds itself by the expansion and development of the cells already formed. It is this phenomenon especially, one altogether peculiar to plants, which, because it depends upon the fact of their being composed of cells, can never occur in any, not even the most remote form in crystals or animals" (p. 251) (translation by Smith [24]).

5. Schwann's cytoblastema

At the time of Schwann's publication of his *Mikroskopische Untersuchungen* [24] the blastema theory was widespread and it would be unfair to charge him with the introduction of it. He is without doubt (see Chapter 6) the one who discovered that animals, as well as plants, are composed only of cells, modified cells or cell products. He certainly made a sad mistake when he extended the blastema theory to animal cells, and this on the basis of slight evidence. He recognized it as

"The fundamental phenomena of all animal and vegetable vegetation".

He states that cells, whether formed inside parent cells or outside these, take place in a structure substance which he calls *cytoblas-*

tema. This was an application of the generally accepted blastema theory to a special case. Though the blastema theory of cell formation known as the cell theory at the time, did not last long, and was soon recognized as erroneous, it remained the basis for the fame of its authors Schleiden and Schwann, whose names were to remain associated.

6. The chemical concept of "molecule" introduced in biogenesis

As shown in Chapter 6 (p. 135) the valuable contribution of Schwann consists in the demonstration that animals are, as well as plants, composed of what he calls the "elementary parts" which are derived, in cell differentiation, from embryonic cells, all organisms being composed of cells, modified cells and products of cells.

One of the features of Schwann's contribution was, as was stressed in Chapter 11 (p. 225), that in Schwann's writing the term "molecule" corresponds not to the "molecule" of the microscopist, but to chemical molecule. We make the point again here because it has been denied [6]. We could quote extensively from the *Mikroskopische Untersuchungen* to justify our conclusion. Suffice it to quote a few lines, written at the time when "elementary atoms" designated our elements or atoms and "atoms of the second order" or "conglomerate molecules", our molecules, and "elementary parts", our cells, including modified cells such as muscle fibres, nerve fibres, etc.:

"As the elementary materials of organic nature are not different from those of the inorganic kingdom, the source of the organic phenomena can only reside in another combination of these materials, whether it be a peculiar mode of union of the elementary atoms to form atoms of the second order, or in the arrangement of these conglomerate molecules when forming either the separate morphological elementary parts of organisms, or an entire organism. We have here to do with the latter question solely, whether the cause of organic phenomena lies in the whole organism, or in its separate elementary parts. If this question can be answered, a further inquiry still remains as to whether the organism or its elementary parts possess this power through the peculiar mode of combination of the conglomerate molecules, or through the mode in which the elementary atoms are united into conglomerate molecules" [24] p. 190).

It should be an extreme sin of "presentism" to recognize in this text a premonition of the molecular order. Within the frame of the ideology of the time, Schwann, leaving the query open, refers to a concept which has remained current in biological thinking (and even until the 20th century), that differences among organisms (for instance, plants and animals) were due to different proportions of the same principles. As shown by Goodman [25], this concept was first formulated by Macquer [26] and stated again by Baumé [27].

7. Plants considered as synthesizers for animals

When the blastema theory was the current scientific belief, the direct nutrition concept was shifted to the molecular level. The "proximate principles" of food were considered as carried to the blastema which was the seat of the formation of cells, in which they were incorporated. As the generally accepted view was that the plants synthesize the nutrients and that the herbivorous animals use these nutrients, the chemical enquiry was principally oriented towards a comparison of corresponding "proximate principles" (For instance, "nitrogenous compounds") in plants or herbivorous animals.

At the time when the method of analysis of natural products consisted in dry distillation, it had repeatedly been observed that, in distillation, animal matters produced more volatile alkali (ammonium), an observation which prepared the way for the concept of the consideration of nitrogen as a characteristic of animal products.

The test of natural alterations was also considered as able to distinguish between animal and vegetable matter. One penetrating remark of Boerhaave [28] is that plant tissues produce alcohol from sugar: an authentic distinction of plants on the basis of alcoholic fermentation.

With respect to distillation, Boerhaave divided the plant kingdom into two sections: the plants which gave acids predominantly in distillation and those which, like animals, gave alkalis (such as *Cruciferae*). This was the first tentative characterization of a section of plant systematics through a chemical approach. But Venel [29] did not verify such distinctions: he also commonly found alkalis in the products of plant distillation, and furthermore he made the sensible remark that treatments by heat modify the "proximate principles" which can only be isolated by a succession of various solvents. This position was likely to discard such claims as that made by Beccari [30] who had called the gluten of flour an animal substance because he considered it as putrefying like an animal corpse.

Berthollet [31], like Venel, was preoccupied with the purpose of comparing "proximate principles". Nevertheless he relied on complex natural products such as wood, skin, tendons, etc. He adopted the test of dissolution by nitric acid, a method first used by Bergmann to dissolve natural substances. Berthollet [31] assumed that the nitrogen produced when natural animal products are treated with concentrated nitric acid at room temperature comes from the animal product, not from nitric acid. In fact, nitric acid contained some nitrous acid which, reacting with the amino groups of α-amino acids, produced nitrogen evolving partly from amino acids and partly from nitrous acid.

Berthollet considered that the plants, in the conditions of the test adopted, did not liberate nitrogen, but only "fixed air" and nitrous gas. These results are, of course, the consequence of a choice of plant materials of predominantly high carbohydrate nature. Animal substances were considered, not as putrefying like animal corpses, but by their constituent nitrogen, a conclusion in harmony with the liberation, in distillation, of *volatile alkali* which had just been identified as ammonia (nitrogen + hydrogen), a logical relation explained by the combination of the nitrogen of animal substances with the hydrogen of water. Animal substances therefore give ammonia in dry distillation. But what of the seeds of the *Cruciferae*? These contain animal substances mixed with their vegetable parts. (Cf. Beccari who had concluded that wheat flour contains a vegetable protein and an animal protein.) Berthollet also considers phosphoric acid as peculiar to animals. But what if Marggraf had isolated phosphorus from plants? To this Berthollet answers that it belongs to the animal parts of the plants. He considers that there are different proportions of animal parts in the matter of plants. This has biochemical implications. For instance, if cows' urine is alkaline it is because the plants they eat contain little animal part and consequently less phosphoric acid. According to Berthollet, plant matter is less complicated than animal matter. The latter contains nitrogen, phosphoric acid and

References p. 85

peculiar oily principles. None of these are found in plants. These distinctions were later weakened and Berthollet [32] finally considered that the differences are quantitative rather than qualitative.

In his *Elements of Chemistry*, Lavoisier [33] follows up Berthollet on this issue. Starting from his concept of organic substances considered as compound radicals joined to oxygen, he considers that the compound radicals of animal substances are made of C, H, N, P and S, while those of plants consist of C and H only. These two elements he considers as common to all plants. Other elements, such as N in *Cruciferae* or P in other plants, were occasional and present only in very small amounts, i.e. inessential (a concept which was to be a source of many errors, although it dispensed with accepting the presence of animal substances in plants).

We have, in Chapter 5 (pp. 117—119), shown the origin and formulation of the concept of "animalization" by Fourcroy and by Hallé which we have presented as the first chemical attempt to define "assimilation", the formation of animal matter, by an addition of nitrogen to vegetable nutrients.

We may add that Fourcroy, using, as Berthollet did, the nitric acid test of Bergmann, found that most nitrogen was liberated by fibrous muscular matter. The albumen matter, known in eggs, but which he had considered having extracted in the form of viscous, heat-coagulable juice of horseradish and other plants (Fourcroy [34]), gave less nitrogen and the gelatinous animal matter from skin or cartilage gave least. This gelatinous animal matter was therefore considered by Fourcroy [35] as approaching vegetable matter, which in his opinion produces little or no nitrogen with nitric acid. These "observations" were the basis of the well-known increasing scale of "animalization": gelatin, albumen, fibrin. This scale was adopted in the classical treatises of Hünefeld and of Richerand (see Chapter 5).

It would be a mistake to believe that Fourcroy saw "animalization" as a mere addition of nitrogen to nitrogenous compounds. In fact, he also recognizes, in this process, changes in the proportions of other elements in "immediate principles". His ideas concerning carbon have an interesting character. While he underlines that carbon accumulates in plants, he points to the fact that it continuously escapes in the respiration of animals, as Black had shown. For him [37], in animals, carbon is a transitory element. This is at the basis of the concept, introduced later and among the first by Dumas, of "respiratory nutrients" of animals as opposed to plastic nutrients. Fourcroy [38] states (vol. 8, p. 257) that plants are chemical instruments which begin the organization of crude inorganic matter, synthesizing it into food for animals.

Goodman [35] has analyzed the tests described by Fourcroy in order to distinguish between animal and plant substance.

But the whole concept of animalization was badly shaken when the student and colleague of Fourcroy, Vauquelin [39], analyzed a plant material which, contrary to the high carbohydrate-containing parts used by his predecessors, was in our language highly proteinaceous; the juice of *Carica papaya*, rich in the enzyme papain, and in which he found much nitrogenous matter.

The animalization theory was a result of the limitation of knowledge and was replaced by a return to direct assimilation according to which the nutrition of animals occurs directly by incorporation of "proximate principles" from plants.

We have in Chapter 6 retraced the introduction by Mulder of the concept of the radical "protein" ($C_{40}H_{62}N_{10}O_{12}$), considered as common to all either vegetable or animal nitrogenous principles. In Chapter 7, we have retraced the introduction by Liebig of the concept according to which plants contain substances whose elementary formula is identical with those of albumin, fibrin or casein. According to this theory, the nitrogenous substances are preformed in plants. As no synthesis is required in animals, fats or sugar play no part in the formation of animal matter.

But there was also a large body of sensualist aspects which were in favor of the concept of direct assimilation of nitrogenous materials. Braconnot [40], for instance, gave the name of *legumin* to a white substance isolated by Einhof [41] from peas and other leguminous plants and which Einhof had recognized as having the odor of cheese. This principle Braconnot [42] later did suggest to be not different from milk casein. A number of publications (literature in Goodman [43]) have emphasized the resemblance of the "milk of almonds" to cow's milk.

Most remarkable was the example of the cow-tree of South America which impressed Humboldt. He suggested to Boussingault, when he visited the regions where it grows, to perform a study of this tree, the milky juice of which clots. This clotted juice was called cheese by the natives.

References p. 85

The myth of protein direct assimilation lasted for a long time, even after the concept of "indirect metabolism" was introduced by Claude Bernard (see Chapter 10). The concept of a chemical biosynthesis of proteins appeared only after it was recognized that proteins were absorbed in the digestive tract in the form of amino acids.

When organic chemists such as Dumas in Paris and Liebig in Giessen plunged into animal chemistry, they enforced the concept of the direct assimilation of nitrogenous matter.

To quote Liebig [44]:

"How beautifully and admirably simple, with the aid of these discoveries, appears the process of nutrition in animals. The formation of their organs, in which vitality chiefly resides! Those vegetable principles, which in animals are used to form blood, contain the chief constituents of blood, fibrin and albumen, already formed, as far as regards their composition. All plants, besides, contain a certain quantity of iron, which reappears in the coloring matter of the blood... Vegetables produce in their organism of the blood of all animals, for the carnivora, in consuming the blood and flesh of the graminivora, consume, strictly speaking, only the vegetable principles which have served for the nutrition of the latter (p. 49).

Animals are qualified by Liebig [44] (p. 47) as "a higher kind of vegetable".

Dumas considers that plants elaborate organic matter, and that animals consume and destroy it. Proteins, fats and sugars of plants were, according to Dumas either retained as constituents persisting in the animal matter, or subject to combustion in the presence of the oxygen of air. In this way, animals (which he compares to steam engines) obtain heat. Combustion became in this context the criterion of animality. He even considered that plants, when they do not receive heat from outside, or when they need heat, behave like animals. They become animals "from the point of view of the general physics of the globe".

REFERENCES

1 F.L. Holmes, J. Hist. Biol., 8 (1975) 135.
2 J.R. Baker, Quart. J. Micr. Sci., 89 (1948) 103.
3 G. Bachelard, La Formation de l'Esprit Scientifique, Paris, 1947.
4 Wallerius, De l'Origine du Monde et de la Terre en particulier (French translation, Warsaw, 1780).
5 J. Hunter, Treatise on the Blood, Inflammation and Gun-shot Wounds, Philadelphia, 1817.
6 L.J. Rather, Addison and the White Corpuscles: and Aspect of Nineteenth-century Biology, London, 1972.
7 J.V. Pickstone, J. Hist. Med., 28 (1973) 336.
8 E. Home, Phil. Trans. Roy. Soc., 108 (1818) 172.
9 P. Mazumdar, Isis, 66 (1975) 242.
10 L. Oken, Lehrbuch der Naturphilosophie, Jena, 1811.
11 F.L. Holmes, Isis, 54 (1963) 50.
12 A. Fourcroy, Mém. Soc. Roy. Méd., (1787) 502.
13 V. Brunner, Der Genfer Arzt Jean-Louis Prévost (1790—1850) und sein Beitrag zur Entwickelungsgeschichte und Physiologie (Zürcher Medizingeschichtliche Abhandlungen, Neue Reihe, N2.41), Zürich, 1966.
14 J.L. Prévost and J.B. Dumas, Bibliothèque Universelle des Sciences, des Lettres et des Arts, Geneva, 17 (1821) 215; 18 (1821) 208.
15 H.M. Edwards, Arch. Gén. Méd., 3 (1823) 166.
16 H.M. Edwards, Répertoire Général d'Anatomie et de Physiologie Pathologique, Vol. 3, first part, 1827, p. 125.
17 J.J. Berzelius, Lehrbuch der Thierchemie (translation by Wöhler), 1831.
18 J. Müller, Handbuch der Physiologie des Menschen, Coblenz, 1833—1840.
19 J. Müller, Ann. Sci. Nat., 27 (1832) 222.
20 J. Berzelius, Jahresber. Fortschr. Phys. Wissensch. (translation by F. Wöhler), 13 (1834) 370.
21 T. Hodking and J.L. Lister, Phil. Mag., 2 (1827) 130.
22 T. Schwann, Tagebuch über Naturwissenschaftliche und Medizinische Beobachtungen und Versuche, Band I (1834—1836) (unpublished).
23 M. Schleiden, Grundzüge der Wissenschaftlichen Botanik, Leipzig, 1842.
24 M. Schleiden, Müller's Arch. Anat. Physiol., (1838) 137. Republished in a collected edition of Schleiden's papers, entitled Beiträge zur Botanik, and translated in English by H. Smith as an appendix to his translation of Schwann's Mikroskopische Untersuchungen. (Microscopical Researches into the Accordance in the Structure and Growth of Animals and Plants, London, 1847).
25 D.C. Goodman, Med. Hist., 15 (1971) 23. (references 26—32 are quoted from Goodman's paper)
26 P.J. Macquer, in A Dictionary of Chemistry (translation by J. Keir), 2 vol., London, 1771, vol. 1, p. 363.
27 A. Baumé, in Encyclopédie Méthodique, Chimie, Pharmacie et Métallurgie, 6 vol., Padua 1786-C 1807, vol. 2, p. 218.

28 H. Boerhaave, Elements of Chemistry (translation by T. Dallowe), 2 vol., London, 1735, vol. 1, p. 375.
29 G.F. Venel, Mém. Math. Phys. Acad. Sci. Inst. France, 2 (1755) 328.
30 I.B. Beccari, Bologna Commentaries, 2 (1745) vol. 2, 122. (On Beccari, see E.F. Beach, J. Hist. Med., 16 (1961) 354.)
31 C.L. Berthollet, Mém. Acad. Sci., (1785) 331.
32 C.L. Berthollet, Eléments de l'Art de la Teinture, 2 vol., Paris, 1791, Vol. 1, p. 131.
33 A.L. Lavoisier, Eléments de Chimie (Elements of Chemistry, translated by R. Kerr, Edinburgh, 1790. p. 145).
34 A. Fourcroy, Ann. Chim., 3 (1789) 252.
35 A. Fourcroy, Encyclopédie Méthodique. Chimie, vol. 4, p. 72.
36 A. Fourcroy, Encyclopédie Méthodique, Chimie, vol. 2, p. 478.
37 A. Fourcroy, Encyclopédie Méthodique, Chimie, vol. 3, p. 64.
38 A. Fourcroy, Système des Connaissances Chimiques. . ., 11 vol., Paris, 1801—1802.
39 L.N. Vauquelin, Ann. Chim. Phys., 43 (1802) 267; 49 (1803—1804) 295.
40 H. Braconnot, Ann. Chim. Phys., 34 (1827) 68.
41 Einhof, Neues Allg. J. Chem., 6 (1806) 115; 6 (1806) 542.
42 H. Braconnot, Ann. Chim. Phys., 43 (1830) 347.
43 D.C. Goodman, Med. Hist., 16 (1972) 113.
44 J. von Liebig, Animal Chemistry or Organic Chemistry in its Application to Physiology and Pathology (translation by W. Gregory), London, 1842 (reprint with an Introduction by F.L. Holmes, New York and London, 1964).

Chapter 42

Animals Recognized as Synthesizers

1. The chemical balance (statics) of organic nature

The concept of the movement of matter in the Universe was familiar to Greek proto-biochemists, for instance from the aspects of the supposed formation of plants from air. In animals, they considered (and this remained a part of the theories of direct assimilation) that all the components of an animal were contained in its food (either vegetable or animal), just as all the components of a tree were contained in air and soil (see Chapter 1). When organic chemistry made progress in the knowledge of organic substances, a number of organic chemists became interested in animal as well as in agricultural chemistry. This approach by organic chemists (and more prominently by Liebig and by Dumas) was directed by a methodology based on a knowledge of organic chemical composition and reactions, on the one hand, and on a knowledge of the net exchange between organisms and environment on the other. No attempt was made to penetrate into the chemical abilities of organisms.

In ancient physiology, the "statical" approach consisted in determining the balance sheet of the material composition of an organism, for instance by weighing or by performing any other quantitative measurement on the organism itself and on its medium.

The classical willow tree experiment of Van Helmont (Chapter 2, p. 75) is an example of such an approach. This trend of research was relied upon by Santorio Santorio (1561—1636) in his famous experiments on insensible perspiration. His example was followed in France by Denys Dodart (1634—1707), physician to Louis XIV (see Grmek [1]), who wrote a *Médecine Statique Gauloise.* As em-

Plate 141. Santorio Santorio.

phasized by Grmek [1], these contributions did introduce quantitative experiments in biological research. They contributed to the transition between the old biological "microcosm-macrocosm" view of Nature and the experimental approach.

In England, Stephen Hales published in 1727 his *Vegetable Staticks* in which he applied the statical methods to plants.

When progress had been in the knowledge of the constitution of organic compounds, static methods were applied to nutrition at large. The idea that the atmosphere constitutes a link between the vegetable and animal kingdoms had been already formulated by Mariotte. The researchers of Boussingault were the determining factor in calling the attention of Dumas to the role of plants in the economy of Nature. Reporting, as a referee, on papers presented by Boussingault to the French Academy of Sciences, Dumas (1838, 1839) expressed his admiration for the manner in which Boussingault introduced organic chemistry in considering the cycles of Nature (Holmes [2]).

In his first report (1838), as Holmes [2] notes, Dumas, referring to the paper by Boussingault concerning plants (see Chapter 40), depicts the significance of these researches in terms of material balance between all organisms. The respiration of animals renders to the air the CO_2 and the water which plants again absorb, supplying the animals with carbon and hydrogen in their food. Dumas gives merit to Boussingault for showing that a similar exchange exists in the case of nitrogen, and showing that plants derive from the air the nitrogen which they provide to herbivorous animals.

In his second report (1839) Dumas states that Boussingault, describing the action of plants on water, air and fertilizers and the modifications of air and aliments by animals, provides, on the basis of accurate chemical analysis, a foundation and a sound basis for agricultural chemistry.

Dumas was prepared to recognize this achievement mainly because of his sincere admiration for Berthollet's *Statique Chimique*. That this book occupied him almost constantly during three or four years, from the age of 17 to 21, is stated by him in his *Leçons sur la Philosophie Chimique* [3] (1837).

In 1838 Dumas had been raised to the important chair of pharmacy and organic chemistry at the Paris Ecole de Médecine. Notes for his lectured there are preserved in the Archives of the Institut

References p. 112

de France (library of the Institut; library of the Academy of Sciences). These unpublished notes have been read by Holmes [2] and from his analysis of these precious documents we may follow the course of the interest raised in Dumas's mind by the work of Boussingault in the fields of the chemical statics of organisms and of agricultural chemistry.

According to the analyses of Holmes [2], it was in 1839 that Dumas placed the chemical phenomena of animal respiration, on which he had previously lectured, within the framework of Nature's equilibria. He stressed that the action of plants on the atmosphere counterbalances the effect of animal respiration. It is at this time (1839) that he formulated the opposition between plants as reduction apparatuses and animals as combustion apparatuses. As we have stated before, the concept of the complementarity of plants and animals was an old one but it was the merit of Dumas to recognize this relation in terms of internal chemical processes, i.e. metabolism. Though he considered the synthetic activities of plants as beyond the possibilities of explanation by organic chemistry, he envisaged the possibility of defining chemical transformations in animals. Those he considered as limited to the action of oxygen operating combustion, mainly in blood.

The main interest of the concept of oxidation of nutrients emphasized by Dumas was that he recognized the existence of "respiratory aliments" (at a time when no relation had been established between food and heat), and that he concluded that food shared with solar light the quality of a source in the flux of energy through organisms. In this perspective, plants become not only a source of plastic constituents but a storehouse of combustible material for animals. He finds in this scheme no place for biogenesis in animals and he divides up nutrients into assimilable products (albumin, fibrin, casein, fats) and combustible products. (Concerning the concept of a distinction between respiratory and assimilable nutrients, the reader is referred to Holmes [2], pp. 38, 41, 56, 184, 267, 416.)

In the last lecture of his course for 1841 (August 20) at the Ecole de Médecine (see Chapter 7) Dumas (a few days after Magendie read the last part of the report of the Gelatin Commission (see below) before the Academy of Sciences, August 2) presented a table contrasting the properties of plants and animals (see Chapter 7, p. 151); plants performing processes of synthesis as the chemist does in his laboratory, while animals obtained their substance by direct assimilation of plastic foods and oxidized "respiratory food" in the blood. Dumas stated that he had conceived this metabolic scheme in constant collaboration with Boussingault.

The lecture was, on the day of its delivery to the medical students (August 20, 1841), printed in a daily newspaper, the widely read *Journal des Débats*, under the following title: *Essai de statique chimique des êtres organisés.* In the fall of the same year it was reprinted in *Annales des Sciences Naturelles*, in which a revised version appeared in December 1841 (a second edition with appendices by Boussingault appeared in 1842 [4] and an English version signed by Dumas and Boussingault appeared in 1844 [5]).

As stated by Holmes [2]

> "Dumas must have been anxious to establish a claim to the ideas he had formulated before it was too late. The cause of that anxiety was undoubtedly the activities of his German counterpart, Justus Liebig".

As was stated in Chapter 7 (p. 149) the *Essai* opposed the chemical properties of plants and animals, though uniting the two kingdoms as "offsprings of the air" and recognizing the atmosphere as the mysterious link connecting them.

2. Liebig and agricultural chemistry

As we have seen, the interest of Dumas for agricultural chemistry was a consequence of reading the papers of Boussingault, whose attention had been called during his stay in South America to the pressing problems of the time with respect to providing food for populations and correcting the resultant situation, all over the world, of the sterility of soils worn out as a consequence of continuous cropping. In France, the years 1828 to 1830 had been disastrous and it had been necessary to replace by potatoes up to 20 per cent of the flour used to make bread.

The origin of Liebig's interest in agricultural chemistry is less clear. When he attended the meeting of the British Association for the Advancement of Sciences in 1837, he was asked to write a review of the state of organic chemistry. Three years later, the book [6] he presented to the Association was discussing agricultural chemistry, though there is no indication that he had been directed to this topic.

References p. 112

M.W. Rossiter [7], in her invaluable book on the emergence of agricultural science, recognizes that, before 1840, Liebig had not displayed the passionate interest he afterwards showed. She suggests different possible sources of this interest. Liebig may have met one or another of those men who were already, in England, actively interested in scientific agriculture *. As one of the papers delivered at the meeting was concerned with agricultural problems, the theory would have some appeal if we did not know (as he wrote home) that on the same day Liebig, who had difficulty in understanding English, had spent his time at the beach. Yet his visit to England in 1837 may have given to Liebig an impetus to deal with agricultural chemistry, as it was in England that scientific agriculture was emerging at the period.

As we have stated in Chapter 7 (p. 145), the interest for agricultural chemistry displayed by Liebig in his book of 1840 was well in line with Lavoisier's trend. As he considered that carbon and nitrogen were provided to plants by aerial nutrition, and that the plants obtained inorganic constitutents from the soil, it was towards these constituents that he directed his attention. On the other hand, when preparing his report in Giessen he was living in the midst of a very poor agricultural area. The high rate of emigration to America among the populations surrounding Giessen may have stimulated his desire to improve agriculture. As emphasized by M.W. Rossiter [7], it is known that it was his visit to Th. Graham, the chemist, during his stay in England, which turned his interest towards the phosphates. We may also consider that on the same occasion he learned, or was reminded of, the fact that fertilizers used in England, besides guano, were partly composed of imported bones from Bavaria. This always hurt his patriotism. But, whatever color these aspects may have given to Liegib's temperamental expressions, the essential stimulation which led him towards writing on agricultural chemistry was undoubtedly, as already pointed out by Lieben (p. 61), his opposition to the humus theory (see Chapter 40).

In the same train of thought, M.W. Rossiter [7] points to the seminal importance of a paper by Saussure [8] published in 1839, and discussing humus.

---

* Probably his former pupil, L. Playfair, who looked after Liebig during his stay in England.

To quote M.W. Rossiter:

"Perhaps Saussure led him to humus, and he then taught himself rapidly all he could about agricultural chemistry. This explanation seems fairly likely, since it would explain Liebig's great interest in the humus in 1849 and his feeling of rediscovering Saussure's earlier work. Fermentation may well have been the topic that brought Liebig and Saussure together" (p. 28).

We have, in Chapter 7, recalled the controversies between Liebig and Dumas after the publication of the latter's *Essai.* In the present section we shall focus our attention on the ability of accomplishing biosynthesis, denied to animals. Both Dumas and Liebig considered that "nitrogenous compounds" were directly incorporated in animal tissues, while Liebig, though considering it an abnormal process (see Chapter 10, p. 202), recognized that animals had the ability to form fats from carbohydrates.

Holmes [2] makes the pertinent remark that from the time of publication of the lecture of Dumas in the issue of August 20, 1841 of the *Journal des Débats,* there "emerged a three-sided contest" with regard to the interaction of biology and chemistry.

There is no doubt that Dumas had formulated before Liebig the concept of "respiratory food", according to which the mutual action, in the blood, of oxygen and of certain nutrients, was the source of animal heat (see Chapter 7, p. 152). While on a different basis, they both believed in the direct intercalation of "nitrogenous compounds", as such, in animal tissues. But they diverged about fat biosynthesis, accepted as taking place in animals by Liebig but denied by Dumas. On the other hand, a contest arose between the "chemists" (Dumas, Liebig and their colleagues) and the physiologists (Magendie, Donné, Bernard, etc.) concerning the method of approaching the problems of nutrition at large (see Chapter 9).

3. The so-called "Gelatin Commission"

At the beginning of the 19th century, the nutritional status of the people had, owing to the poor returns of agricultural processes, become a source of great concern. In France, great hopes had been put on the feeding of the poor with bouillon made from the gelatin of bones. The "gelatin soup" was even mentioned in the foreword, written in 1834 by Th. Gautier, of his famous novel

Plate 142. François Magendie.

Mademoiselle de Maupin, along with continuous-flow syringes or railways as examples of utilitarian devices which he distinguished from the forms of beauty in literature, such as a book, a novel or a sonnet.

Holmes [2] (p. 9—13) has written an excellent history of the Commission and the reader is referred to this detailed account.

The conclusion of Magendie (1816), according to which animals must receive nitrogen with their nutrients, increased the confidence in gelatin, a nitrogenous compound, supposed to complete a diet of predominantly vegetable nature.

After gelatin soup had been fed to hospital patients and to the workers in a number of workshops, a skeptical reaction came, in 1835, in the form of a pamphlet written by the clinician Donné [9]. From this time on, many maintained that no proof was available of the nutritional value of gelatin. Several authors even suggested that it might be harmful. A Commission was appointed to evaluate Donné's paper. It was chaired by Thenard and the members were Chevreul, Dumas and d'Arcet (a chemist who had devised a process for the extraction of gelatin from bones). It soon faced an increasingly heavy task. In 1836, Magendie undertook for the Commission a series of experiments on dogs in which his disciple Bernard helped him.

Magendie [10], presenting the report of the Commission to the Academy of Sciences in 1841, concluded that gelatin was not an adequate nutrient. Liebig did not take kindly to this. That his argument was based on his simple form of chemical "intercalarism" appears in a letter to Pelouze in which [11] he writes:

> ". . . to conclude from the facts which one has observed that gelatin cannot serve for something in the organism is an absurdity. It cannot replace or be changed into blood, nor make fat, but it can serve for the reproduction of membranes and of the substance of cells, and it does serve for that" (translation by Holmes [2]).

It was shown much later that the inadequacy of gelatin as a nutrient lies in the fact that it is deficient in the essential amino acid tryptophan.

It was shortly after the presentation (July 5 and 26; August 2) of Magendie's report that Dumas had his *Essai* published by the speedy medium of a daily newspaper (August 2). Thus were the outposts set up for the war on the biosynthesis of fats which was soon to rage.

References p. 112

4. The controversy between chemists concerning the ability of animals to synthesize fats

We have, in Chapter 10 (p. 202), quoted from Liebig's *Animal Chemistry* a sentence in which he considers (at the time of writing this book) the production of fats by animals. The process is considered by him as a consequence of a deficient supply of oxygen. This process, he writes, takes place "under the well known conditions of the fattening of domestic animals". It was in fact a widespread popular concept, which had been considered as an example of transmutation (for instance by Ingen-Housz, see Chapter 40).

Dumas and his school, sticking to their grand scheme of the complementarity of plants and animals, and to the notion that at the time no conversion of any substance into fat had been accomplished in the laboratory, did not consider as an argument the piece of "writing-desk chemistry" which Liebig had produced to explain this conversion. They believed that animals obtained their fat from their food.

In his book on Bernard (Chapter III—V), Holmes [2] has written an exhaustive and documented history of the polemics on fat biosynthesis, based on a rich collection of manuscript sources, and his contribution to an analysis of the attitudes of the scientific leaders of the time deserves the highest praise. In the remaining part of this Chapter, we shall epitomize the text of Holmes [2] and draw on it with respect to the analysis and bibliography of the primary literature and unpublished sources he has perused. This does not dispense the reader from referring to the detailed treatise by Holmes.

As we have stated above, Liebig's contention was based on common knowledge and on a calculation. As noted by Holmes, Liebig "no less than Dumas, was engaged, in considering the problem of nutrition, in speculating about the processes supposed to go on inside animals with no direct knowledge of them", a position which was to be opposed by physiologists and most successfully by Claude Bernard in his work on the glycogenic function of the liver, in which the physiological reality concerned was discovered by experiments on animals and later confirmed by chemical observations in vitro (see Chapter 10).

The first paper on animal chemistry by Liebig, translated in French by his friend Pelouze, appeared in the *Annales de Chimie*

[12] with two editorial critical notes, the second of which read as follows:

"But M. Liebig thinks that herbivorous animals *make* fat with sugar or starch, whereas MM. Dumas and Boussingault assert as a general rule that animals, of whatever kind, *do not make* fat or any other organic alimentary material, and that they borrow all of their aliments, whatever they are sugar, starch, fat or nitrogenous substances, from the vegetable kingdom. If the assertion of M. Liebig were well-founded, the general formula stated by M.M. Dumas and Boussingault as a summary of the *Chemical Statics* of the two kingdoms would be false.

"But the Gelatin Commission has placed beyond doubt that animals which eat fat are the only ones where one finds that fats accumulate in the tissue" (translation by Holmes [2]). [This note was signed *R* (for Rédacteur).]

It was true that the Commission had noted that animals fed for long periods on fat alone had abnormally large quantities of fat infiltrated through their tissues, but as Holmes [2] notes, there was no conclusion in the report written by Magendie that animals cannot convert other substances into fat. On the other hand, the note does not claim that there was such a conclusion.

In his *Animal Chemistry* Liebig [14] repeated his argument in favor of the conversion of starch and sugar to fat. As Holmes [2] states, these reasons were rather slim.

"... they consisted of the common knowledge that domestic animals are fattened by feeding them grains and the speculation that the chemical reaction would require only the removal of oxygen from the sugar molecule".

In his *Animal Chemistry* [14] Liebig added two more reasons to strengthen his view. Gundlach [13] had shown that bees can produce wax on a diet of honey. Liebig reproduced portions of Gundlach's book in an appendix to his own book. On the other hand, he wrote in the preface:

"When a lean goose, weighing 4 lbs, gains, in thirty-six days, during which it has been fed with 24 lbs of maize, 5 lbs in weight, and yields $3\frac{1}{2}$ lbs of pure fat, this fat cannot have been contained in the food, ready formed, because maize does not contain the thousandth part of its weight in fat or in any substance resembling it".

To quote Holmes [2],

"As he later explained, while he was writing his preface, a friend had informed him of the results of fattening a goose. He had then simply looked up in the chemical literature two different analyses of maize which made no mention that it contained any fats or oils. Without making a special investigation himself, he inserted his conclusion into his introductory discussion of methods, too hastily apparently to revise the main treatment of the topic in his book" (p. 57).

Dumas and Payen reported in a note presented on October 24 to the Academy that they had found that maize contains about 9 per cent of oil. From this and from Liebig's figures they calculated that the food could have supplied 1.25 kg. As the goose had yielded 1.75 kg of fat, the remaining amount presumably existed in the goose prior to feeding.

The position of the French chemists appears here in the light of their methodology. They based their denial of a synthesis of fats in animals on their conviction that the food provided the amount of fat. The whole issue was to concentrate on this statical aspect (balance sheet). This methodology of the organic chemists contrasted with the approach of Claude Bernard who proved the biosynthesis of glycogen in animals by utilizing vivisection methods inside the body.

Liebig [15], who had first been led essentially by common knowledge of fattening pigs with grain, turned to an analysis of evidence provided by Boussingault himself. Boussingault had on the basis of a consideration of the chemical elements, made measurements of the intake of food on the one hand, and of the production of urine, excrements and milk of a cow during several days in order to see if the (potato and hay) intake, particularly of nitrogen, balanced the losses (see Holmes [16]).

Analyzing himself the content of fat in the food given to the cow, Liebig concluded that the variation of these foods in fat content was too great to lead to any conclusion. On the other hand, he provided data from which he concluded that the intake of fat in the diet seemed inadequate to account for the fat obtained. He raised objections concerning the analytical methods used by Dumas and his school, in considering the ether extract as giving the content in fat, and made the pertinent remark that the ether extract of potato or hay included chlorophyll and other compounds and did not contain the fatty acids of butter or of animal fat. On the other hand, he showed that waxy compounds contained in the cow's food were nearly all recovered in the feces and could not contribute to fat formation. In the same paper, Liebig expressed his agreement with Dumas that animals cannot synthesize the nitrogenous substances but he disagreed with the general opposition between animals and plants accepted by Dumas. This disagreement he expressed by stating [15]

> "that with respect to the formation of many of their constituents, similar processes can take place in animals as do in plants" (translation by Holmes [2]).

At the time, Boussingault was experimenting on his farm on practical livestock raising and he became more interested in the fat problem, and in the "fatty and waxy principles of hay", when he observed that cows did not do as well when fed entirely on roots, as when the usual hay and wheat were added. This appeared to reinforce the position of Dumas based on general principles, and thus Boussingault, on a practical basis, supported this position.

5. Payen joins with Dumas

We have recalled in Chapter 13 (p. 267) the important role played by the discovery by Payen and Persoz of what they called *diastase* as the starting point of the enzymatic theory of metabolism. Payen became interested in the problem of fat biosynthesis on a practical basis, as had been the case with Boussingault. But he was also inclined to adhere to the sharp contrast claimed by Dumas to exist between animals and plants. The conceptual basis of his strong support came from the fact that Payen had been closely interested in the composition of plant tissues and had shown that the unity of the plant kingdom was illustrated by the general presence, in plant "membranes", of a common and uniform compound, cellulose, which he had discovered. On the other hand, Payen, by his extensive analytical studies, had convinced himself of the universal distribution of fats among plants. In 1843, the cooperative effort of Dumas, Boussingault and Payen materialized in a common publication in which they compared the fatty material of food and the amounts produced by the animals, concluding that a cow extracts from its nutrients nearly all of the fatty material which they contain, and that this fatty material is converted into butter.

References p. 112

6. Recurrence of nitrogen as a testimonium of animal nature

Concerning the discovery, by Payen, of the general presence of cellulose in plant cells, we may note here that he also claimed that animal cells are bounded by a quaternary, nitrogenous membrane [18]. He recognized in this issue a distinctive property of plant membranes. Nägeli [19] adopted the distinction and considered that the nitrogenous boundary of animal cells accounts for sensation and motion, lacking in plants. This criterion was accepted by Robin [20] (part II, p. 186). The principle was to be invalidated by the discovery of animal cellulose in tunicates accomplished by Schmidt [21] (p. 61) and confirmed by Loewig and Kölliker [22].

7. Plant fats and animal fats compared

As Liebig had rightly emphasized, the animal fats are chemically distinct from plant fats. The French chemists agreed that the fatty matter is chemically modified in the animal but cannot "se former de toutes pièces". But they also maintained that all transformations in animals were oxidative transformations. Waxes of plants were, according to their view, converted by a "commencement" of oxidation into stearic or oleic acid.

To quote Holmes [2]:

> "Thus their general statements seemed to be moving towards Liebig's view that conversions, if not the creation, of nutritive materials take place in animals, but substantively they still opposed his position by admitting only one class of conversion" (p. 64).

As Liebig saw that they had not taken into account his recalculation of Boussingault's data, in a letter to the Academy [23], read at the session of March 6, he repeated his calculation. He added that he had observed that the excrements contain almost all the waxes ingested by a cow. He also made the remark that it was difficult to conceive how it was possible that

> "a substance which is not saponifiable, and whose melting point is higher than the temperature of an animal, can pass into its blood to undergo there oxidation and transformation into stearic acid" (translation by Holmes [2]).

During the debate on Liebig's letter, Dumas presented arguments which he wrote up in a text presented during the Academy meeting of March 13, 1843. He pointed out that the data taken by Liebig from Boussingault's publication concerned two different cows and that therefore his conclusions concerned "animals too chimerical" to be accepted. He also wondered how Liebig, who could quite well imagine that fibrin, sugar, etc. could be converted into fat, could not conceive this for waxes. Nevertheless he made no answer to the pertinent remarks of Liebig concerning the intestinal absorption of wax. Liebig sent an answer which was read at the meeting of April 3, 1843. He claimed that his experiments were not fictitious, for he had measured the amounts he mentioned on a cow at Giessen. It turned out that this cow produced the same amount and ate the same amount as the cow in Boussingault's experiment [24].

To this Dumas [25] answered that as the figures mentioned by Liebig were the duplicates of those published by Boussingault, it was not credible that Liebig had obtained them in independent experiments. Dumas returned to the objections raised by Liebig concerning the conversion of wax to animal fats and quoted from recent work of Lewy in Copenhagen. This author had shown that waxes were saponifiable when treated with boiling potash. On the other hand, the elementary composition was, according to him, the same: $C^{68}H^{68}O^{4}$. This favored the contention of Dumas that waxes could be converted into fatty acid by a small degree of oxidation.

Dumas concluded that only quantitative feeding experiments could lead to a solution. Boussingault [26] spoke after Dumas. From a balance sheet experiment he concluded that 1614 g of fatty matter had been provided in the food, and that the total in milk and excrements amounted to 1413 g. Commenting on this result, taken as confirming his views, Dumas remarked that (as he put it politely), "if the experiment of M. Liebig is real", such carefully controlled experiments as those of Boussingault appear nevertheless to dominate the discussion.

8. Emphasis shifted from the opposition of animals and plants to possible precursors of fats in animals

To the letter of Liebig read at the meeting of the Académie des Sciences on April 3, 1843, Payen [27] answered during the ses-

References p. 112

sion which took place on April 17. Liebig had remarked that the food eaten by cows does not contain butter. To this Payen answered that he, and his French colleagues, had always said that nutrient fats were fixed in a more or less modified state. They had always included under the denomination of fatty materials such substances as oils, waxes, etc.

"because of the evident analogy which exists in their elementary composition. It is always these materials, according to us, to which the role falls which M. Liebig attributes to starch and to sugars" (translation by Holmes [2], p. 68).

This is a most interesting text. As Holmes [2] acutely remarks when commenting on it, the problem has shifted from the general types of natural processes (reminiscent of the old macrocosm-microcosm view of Nature) to the consideration of possible chemical precursors involved in a chemical synthesis.

As also noted by Holmes [2], it is clear that the French chemists would not have provided the proof of the capability of animals to biosynthesize fats from sugars if they had not been prodded by Liebig

"to clarify specific aspects of their scheme at the expense of its overall simplicity" (Holmes [2], p. 68).

It certainly was a notable development that Dumas, Boussingault and Payen [28], during the same year 1843, considered possible ways of chemical transformation of sugar into animal fats. This was the first historical endeavor of guessing, on a chemical basis, about possible pathways of biosynthesis. The authors represent sugars as composed of CO, water and ethylene. Since ethylene can be converted into various alcohols yielding fatty acids upon oxidation there is

"chemically speaking, nothing to oppose the possibility that in digestion sugar gives rise to oils which take part in the formation of fats" (translation by Holmes [2], p. 68).

The penetrating analysis of Holmes [2] leads to interesting aspects of the methodology of the chemists in their approach to physiological problems. Dumas and Liebig shared the same criteria but differed in their application. For instance, they agreed on their application of Occam's razor (see Chapter 10, p. 180), assuming as they both did that that theory which required the least complicated chemical transformations was to be preferred.

To quote Holmes [2]:

"Liebig considered his explanation of the conversion of sugar to fat convincing because it involved only a removal of oxygen. He questioned the conversion of wax to fat at least partially on the ground that wax was chemically quite different from fat and consequently the conversion would be complex. Dumas, on the other hand, argued in favor of the transformation of wax to fat in that it required a less extensive chemical alteration than the changes one would have to imagine if sugars, starches, and other substances served as sources. The French chemists used, as a reason for advocating the nutritive sources of all animal fat rather than an alternative theory, the argument that "it is the simplest way to explain it". So long as nutritive fat was shown to provide an adequate supply, they said, "nothing authorizes (us) to regard the animal as capable of producing "fat" [29]". (p. 69)

One common feature of the methodology of the chemists was the ability to explain a transformation of substances supposed to occur in an organism on the basis of the chemical composition of the compounds involved and on the basis of an equation. A fully satisfactory verification required, in addition, the imitation of the proposed reaction in the chemical laboratory.

In contrast with the requirement of such methodological conditions as just stated was the acceptance by the chemists of the biosynthesis they recognized as accomplished by plants, an axiom of their statical system.

As Holmes [2] acutely remarks:

"Neither seemed bothered by the fact that they accepted that plants somehow can perform all of the synthetic reactions which were then beyond any possibility of artificial duplication" (p. 70).

9. Objections of the physiologists

To requests that they satisfy not only their methodological criteria, but also those presented by the physiologists, Dumas and Liebig were relatively impervious. Their myopic views in this respect are exemplified in their attitude towards such an eminent disputer as Magendie. When Liebig, in a letter to the Académie, stated that waxes eaten by an animal are excreted in its feces, Magendie declared his agreement with Liebig's views. But two weeks later he was compelled to recognize that he had made a mistake in his reasoning. This admission did not prevent him from remarking

References p. 112

that it certainly was important to know that compounds are found in plants which are similar to the organic constituents of animals, but that it is a big step from there to demonstrate that the tissues of animals are exclusively formed from those vegetable materials. And he added that in his opinion this bridge could only be crossed by direct experiments on animals. This discussion was reported by Donné [30] in the *Journal des Débats.* Did Dumas, he wrote, believe that serpents have venoms because they eat venomous substances? It is interesting to find in this report a statement which formulates the current view of the time regarding metabolism as conceived by the "analysts". According to this theory, the animals receive from their nutrients only the elements entering in the composition of the substances which they need.

"They rejoin these elements, associate them, combine them to make from them first the blood, and later, by a marvelous workmanship of the organism, by a series of elaborations and transformations, fibrin, fat, albumin and finally flesh, bones, nervous tissues, seminal fluid, bile, milk, urine, etc." (Donné [30], quoted after Holmes [2], p. 74).

But, as Holmes notes, the first shock to the position of Dumas which influenced his views did not come from the physiologists, but from the discovery of new chemical facts.

10. Discovery of a transformation, by organisms, from one category of chemical compounds to another category

Boutron and Frémy [31] showed in 1841 that when sugar, under carefully controlled conditions, is fermented with animal matter, it is completely transformed into lactic acid. This observation belongs to the introduction of glycolytic studies on animal tissues which were reported in Chapter 20. But, in the present context, these observations are important in leading to their repetition, in other conditions, by Pelouze and Gélis [32], who obtained butyric acid identical with that derived from butter. From their careful analyses, Pelouze and Gélis [32] were led to revise the empirical formula of butyric acid and to propose a chemical equation:

$$\overbrace{C_{12}H_{28}O_{14}}^{\text{glucose}} = \overbrace{C_8H_{14}O,\ H_2O}^{\text{butyric acid}} + 4\,(CO_2) + 8\,H + 2\,(H_2O)$$

They also obtained from butyric acid and glycerol an oil which they tentatively identified with a component of butter, butyrine, discovered by Chevreul.

This very important work was the basis for the development of the chemical synthesis of fats by Berthelot, a student of Balard at the Collège de France, who had first worked with Pelouze. But, in the present context, the breakthrough consisted in the conversion of a sugar into a fat, a metamorphosis so far unknown. Its importance appeared quite clearly to Pelouze and Gélis [32], who wrote:

"This observation will inevitably occupy an important place in the present discussion of the formation of fats in animals. Without wishing in any way to judge in advance the methods which Nature employs in the numerous modifications which she causes aliments to undergo, we cannot refrain from remarking that the transformation of sugar into butyric acid is carried out without the intermediary of any large increase of temperature and without the application of any of those energetic reagents which would be apt to destroy the equilibrium and the vitality of the animal economy; but that on the contrary the transformation takes place under very simple conditions and with materials which organic nature itself makes available" (translation by Holmes [2], p. 80).

They added that other members of the series of animal fatty acids might have a similar relation to sugar and starches.

11. Reaction of Payen, Dumas and Liebig

In the discussion of the communication by Pelouze at the Académie des Sciences, Payen recalled that Dumas, Boussingault and Payen had already concluded that, chemically speaking, nothing was opposed to a conversion of sugar into fat. But of course, the observation of the phenomenon would bring about a new point of departure along a course which would run counter to that of Dumas and his group. This aspect probably seemed logical to Dumas, who remained silent during the discussion. Liebig's reaction appeared a few weeks later, in a letter to Dieffenbach, as follows:

"In (the question of) the formation of fat I have carried off a glorious victory. Pelouze has found that sugar in contact with putrefying and fermenting matter decomposes into butyric acid and hydrogen gas" (translated from manuscript by Holmes [2], p. 82).

References p. 112

Plate 143. Théophile Jules Pelouze.

12. The sources of bees' wax

As they considered that the arguments proposed by Liebig in favor of a biosynthesis of fats by animals were subject to difficulties of interpretation, H.M. Edwards * and Dumas thought of studying the well-known production of wax by bees. The blind Swiss naturalist, Huber, had from 1789 to 1800, with the help of an intelligent servant, performed a series of most interesting observations on bees, published in 1814 in a book [33] which has remained a classic of entomological literature. In this book, Huber reported that the bees produce wax even if entirely deprived of fats in their food. Huber observed that when bees are kept inside the hive and fed with honey or sugar, to the exclusion of any other kind of nutrient, they continue to produce wax in large amounts. This observation had been confirmed by Gundlach [13].

Concerning this observation, Payen at a meeting of the Académie des Sciences held in February 1843, had maintained that the bees formed wax from fatty material already stored in their body. In August, Edwards and Dumas reported on balance experiments, taking into account the pre-existing fatty material in the body of the bees. Their conclusion, which was developed in an extensive paper, was that the bees actually produce wax. It appears therefore that Dumas and Edwards accepted the verdict of factual evidence.

As Holmes [2] adequately remarks.

"Real scientists, however, often do not read as logic would have it" (p. 84).

This aspect appears in the discussion at the Académie des Sciences. During this exchange of views, Payen redefined the thesis of the French chemists.

In the nutritional issue at hand, he recognized three different problems. The first two problems consisted in asking whether vegetable substances contained fatty matter, and whether these fatty matters were assimilated by animals. The French chemists had from the start of the discussions proposed these concepts and the evidence accumulated was in favor of it. Another question is whether substances belonging to other chemical categories can

---

* Often incorrectly called H. Milne-Edwards, as was noted in Chapter 41.

References p. 112

form fats in animals. Payen distinguishes two aspects of this query. The first one is the transformation of albumin, fibrin or gelatin into fat. At the time neither Liebig nor Dumas accepted this. The second aspect was the transformation of sugars into fatty matter. Concerning this point, Payen made it sound as if the French chemists had always urged the possibility of the process but had cautioned that the evidence was still lacking. On the other hand, Payen saw nothing in common between specialized secretion, such as the production of wax, and the fattening of pigs. Edwards' answer to Payen points to the flaw persisting, in spite of the evidence, in the conceptual system of the French chemists who had not yet developed an interest in the existence of a metabolic pathway of biosynthesis per se, but continued to believe in the standard of simplicity. If dietary fat accounts for the major part of the fatty products of animals there is no interest in considering the formation of fat within the animal organism. In answering Payen, Edwards agreed that foods containing fats were more effective in livestock fattening than were non-fat foods and that this would continue to be so, even if it were shown that all animals could convert sugar to fat. Edwards even neglected the evidence brought up by Pelouze when he formulated his metabolic theory, sticking to the consideration of chemical transformations kept, within organisms, inside each category of compounds. He [36] agreed that the facts recorded concerning bees were not of importance in the debate on fat synthesis by animals. He stated that it appeared to him easy to admit

> "that albumin can become fibrin or gelatin, that the oil of almond can become animal fats. Each aliment thus serves only for the creation of a certain order of products, and consequently the aliments of different families cannot replace each other in their physiological roles. Envisioned from that point of view, the discussion which the Academy had witnessed for the last several months will only be secondarily influenced by the question raised in our note" (translation by Holmes [2], p. 87).

13. Observations on geese by Persoz

In October 1843, two more observations favorable to the ability of animals to synthesize fats were presented to the Académie des Sciences. Chossat [37] reported on experiments made by him several years before, showing that animals fed nothing but sugar until they died still contained nearly normal deposits, while those who died as a consequence of complete abstinence contained almost none. On the other hand, Vogel [38], considering the generally attested fattening effect of beer, found that beer contains only small amounts of fatty matter.

But the most impressive data were announced by Persoz [39] at the Académie on February 12, 1844. We recalled in Chaper 13 the participation of Persoz with Payen in the discovery of diastase. He had, in 1837, become professor of chemistry at the University of Strasbourg where he had the opportunity of becoming acquainted with the Alsatian method of producing fatty liver in geese.

In 1838, Persoz had observed that geese, treated in the Alsatian manner, gained a greater proportion of fat than their proportional increase in weight. He therefore suspected that substances of another chemical nature had been utilized in the production of fat. When the production of butyric acid was demonstrated by Pelouze, and when the experiments on the production of wax by bees were known, Persoz decided to repeat his experiments in the light of his recognition of a possible origin of fat in substances of another chemical category. He experimented on ten geese, which he killed in succession after periods of approximately twenty days. He compared the gain of weight, of fats and the amounts of fatty materials received by the animals with maize, and he concluded that a large proportion of the fat had formed at the expense of the sugar and the starch of the maize.

14. Experiments by Boussingault on pigs

In the spring of 1844, the French school still maintained that animals find their own substances in the aliments which nourish them, as was shown by numerous texts (see Holmes [2], Chapter V). But the central issue, announced by the experiments of Persoz on geese, was finally settled by the results obtained on pigs by Boussingault [40,41]. These results, which appeared in an extensive paper in 1845, had been announced to the Académie des Sciences by a letter from Boussingault read during the meeting of June 16, 1845. He confirmed the results obtained by Persoz on geese. Among the results he obtained on pigs, Boussingault showed

References p. 112

that in the fattening of pigs there is much more fat assimilated than is found in their rations. This was a clear verdict in favor of Liebig's position, though his name was not mentioned. It is clear that the demonstration of this position was provided by Boussingault in his classical work on pigs.

After the letter was read, Edwards stated that the results were in accordance with the results obtained by Dumas and Edwards on bees. Payen maintained that, in spite of the nature of the physiological issue the practical aspect of the French theory had been maintained (see Holmes [2], p. 104).

Liebig [42] pronounced, quite incorrectly, in what is characterized by Holmes [2] (p. 105) as "a display of arrogance", that none of the results of Boussingault added the slightest amount to common knowledge acquired by agricultural practice. He also, this time quite fittingly, scorned Payen for now acting as though his former views had not existed.

Now that the issue had been settled by Boussingault, through long and well-criticized experiments, in favor of what had been, on the part of Liebig, an idea based on popular knowledge, it may be concluded that Liebig's contribution had been no more than a catalyst to spur on the French chemists.

But, as Holmes states:

> "Neither side would have done what it did without the stimulus of the other, so that the outcome was a product of their indissoluble net of interaction" (p. 108).

One important point made by Holmes (p. 108) concerns the influence of the controversies reported in this Chapter on the concepts of Liebig and of Dumas, who, in spite of seeing the problems through the limited perspective of the chemist, were men of different stature than Payen or Edwards.

Both Dumas and Liebig, by 1845, realized the difficulties of enquiring about what happens within the animal organism and the unreliability of the accomplishment of chemical reactions in the laboratory as a proof of their natural occurrence. But, at the same time,

> "they justified their assertion that their views of animal nutrition comprised a suitable framework within which to pose and answer more specific questions" (Holmes [2], p. 108).

It is historically clear that the demonstration of the power of biosynthesis in animals is due to Boussingault, who provided the proofs of what had previously only been an idea. But this accomplishment (by the statical method) was soon overshadowed by the discovery by Claude Bernard (by the experimental method) of the glycogenic function of the liver. (This was reported in Chapter 10, in Part II, which was devoted to the main epistemological changes which have contributed, along a period of time, to the conversion of proto-biochemistry into biochemistry.)

When closing the section on his book on Claude Bernard, devoted to the controversy on fat synthesis, of which the present chapter epitomizes the rich collection of data and analyses to which we refer the reader, Holmes [2] acutely remarks:

"Ironically, Bernard shortly afterward utilized a discovery based on an investigation quite similar to that of Boussingault, but involving sugar in the blood rather than fat, to illustrate the shortcomings of the kind of chemical physiology with which Boussingault had been identified" (p. 117).

REFERENCES

1 M.D. Grmek, Introduction de l'Expérience Quantitative dans les Sciences Biologiques, Paris, 1962.
2 F.G. Holmes, Claude Bernard and Animal Chemistry. The Emergence of a Scientist, Cambridge, Mass., 1974.
3 J.B. Dumas, Leçons sur la Philosophie Chimique, Paris, 1837.
4 J.B. Dumas, Essai de Statique Chimique des Etres Organisés, 2nd ed., Paris, 1842 (with appendices by Boussingault).
5 J.B. Dumas and J.B. Boussingault, The Chemical and Physiological Balance of Organic Nature: an Essay (misprinted as Cssay), New York, 1844.
6 J. Liebig, Organic Chemistry in its Applications to Agriculture and Physiology (translation by L. Playfair), London, 1840.
7 M.W. Rossiter, The Emergence of Agricultural Science, Justus Liebig and the Americans, 1840—1880, New Haven and London, 1975.
8 Th. de Saussure, Bibl. Univ., 13 (1838) 380.
9 A. Donné, Mémoire sur l'Emploi de la Gélatine comme Substance Alimentaire, Paris, 1835.
10 F. Magendie, Compt. Rend., 13 (1841) 276, 280.
11 Letter of Liebig to Pelouze, Nov. 1, 1841. Dossier Dumas, A., Académie des Sciences (quoted from Holmes [2], p. 35).
12 J. Liebig, Ann. Chim. Phys., 4 (1842) 187.
13 F.W. Gundlach, Die Naturgeschichte der Honigbienen, Cassel, 1842.
14 J. Liebig, Animal Chemistry or Organic Chemistry in its Application to Physiology and Pathology (translation by W. Gregory), London, 1842.
15 J. Liebig, Ann. Chem. Pharm., 45 (1842) 113.
16 F.G. Holmes, Isis, 54 (1963) 71.
17 J.B. Dumas, J.B. Boussingault and A. Payen, Compt. Rend., 16 (1843) 348, 352.
18 A. Payen, Compt. Rend., 17 (1843) 227.
19 C. Nägeli, Z. Wiss. Bot., (1845) 1.
20 C. Robin, Du Microscope et des Injections, Paris, 1843.
21 C. Schmidt, Zur vergleichende Physiologie der Wirbellosen Thiere, Braunschweig, 1845.
22 C. Loewig and A. Kölliker, Ann. Sci. Nat. (Zool.), 5 (1846) 193.
23 J.Liebig, Compt. Rend., 16 (1843) 552.
24 J. Liebig, Compt. Rend., 16 (1843) 664.
25 J.B. Dumas, Compt. Rend., 16 (1843) 673.
26 J.B. Boussingault, Compt. Rend., 16 (1843) 668.
27 A. Payen, Compt. Rend., 16 (1843) 770.
28 J.B. Dumas, J.B. Boussingault and A. Payen, Ann. Chim. Phys., 8 (1843) 63.
29 J.B. Dumas, J.B. Boussingault and A. Payen, Compt. Rend., 16 (1843) 353.
30 A. Donné, Journal des Débats, October 26, 1842.
31 Boutron and E. Frémy, Ann. Chim. Phys., 2 (1841) 257.

32 J. Pelouze and A. Gélis, Compt. Rend., 16 (1843) 1263.
33 F. Huber, Nouvelles Observations sur les Abeilles, Geneva, 1814.
34 J.B. Dumas and H.M. Edwards, Compt. Rend., 17 (1843) 531.
35 J.B. Dumas and H.M. Edwards, Ann. Sci. Nat., 20 (1843) 174.
36 H.M. Edwards, Compt. Rend., 17 (1843) 545.
37 Ch. Chossat, Compt. Rend., 17 (1843) 805.
38 Vogel, J. Pharm. Chim., 14 (1843) 309.
39 J.F. Persoz, Compt. Rend., 18 (1844) 245.
40 J.B. Boussingault, Compt. Rend., 20 (1845) 1726.
41 J.B. Boussingault, Ann. Chim. Phys., 14 (1845) 419.
42 J. Liebig, Ann. Chem. Pharm., 54 (1845) 376.

Chapter 43

Aspects, in the Field of Biosynthesis, of the Theory of "Protoplasm"

1. "Living albuminoids"

In the very early biochemical literature, the concept of living molecules (particles) has repeatedly appeared, for instance in the theories of Buffon (see Chapter 4) or in the concept of "living fibre", individually alive, of Diderot [1]. One of the theories belonging to that category has been Huxley's theory of protoplasm formulated in 1869 (see *Introduction* to Part III). In this theory, a concept akin to the notion of transmutation is postulated. In transmutation as understood in several forms of proto-biochemistry (see Part I), the concept of the transmutation of one of the classical four elements into another remained a philosophical notion. Lavoisier has defined the chemical elements and recognized their persistence, as well as the fact that plants and animals were constituted of the same chemical elements. Transmutation into protoplasm was supposed to result from the incorporation of a biomolecule into the protoplasmic complex. In this form of assumed biosynthesis, from dead matter, living matter was obtained by introduction into protoplasm. It was also recognized that the "albuminoids" (later called proteins) were the most important constituents of "protoplasm". The problem therefore was to find out, as Fruton [2] formulates in a penetrating essay on "Energy-rich proteins, 1870—1910":

"How does the molecular motion of intracellular proteins lead to the metabolic transformation of food materials into protoplasm ("vital synthesis") and to the energy changes evident in such biological phenomena as the production of heat, of muscular contraction, of bioluminescence?" (p. 18).

We have, in Chapter 26, discussed the aspects of Pflüger's theory concerned with the energy sources in organisms, based on energy-rich cyanogen bonds. We refer the reader to the penetrating analysis of Fruton [2] concerning a number of variants on that theme. It is on the synthesis of these assumed energy-rich proteins that we shall concentrate in the following sections which are greatly indebted to Fruton's study.

2. Synthesis of "living albuminoid" ("lebendiges Eiweiss") from alimentary albuminoid ("Nahrungseiweiss"), according to Pflüger

In his classical paper of 1875 [3] (p. 300) Pflüger refers to "alimentary albuminoid" as modified by incorporation into protoplasm, and losing its indifference to oxygen. Becoming "living albuminoid", it begins to live. For Pflüger, the energy was released by the dissociation of a labile group which he identified as cyanogen. In modern language, Pflüger recognizes the presence, in cyan proteins, of a high-energy bond, detected by thermochemical measurements, and capable of splitting with a development of energy, i.e. what Hermann had called an energy-generating or "inogen" substance (see D.M. Needham [4], p. 37). Pflüger recognized the alleged biosynthesis of the protoplasm as an endergonic process. It is also in accordance with the protoplasm theory that Pflüger considers cyan proteins ("lebendiges Eiweiss") as endowed with the polymerization faculties of the cyan derivatives, with a preference towards the adjunction of units of the same nature and growing to enormous polymers, which are themselves again dissociated. In this context Pflüger points to the fact that his living albuminoids do not have any constant molecular weight. In short, Pflüger identified life with the intramolecular heat of very labile units of the protoplasm, these "living albuminoids" or high-energy units being regenerated, and growing by polymerization to large aggregates.

Pflüger considers his "living albuminoids" as possessing a chemical configuration different from that of the "dead albuminoids" available for analysis by the chemist. While the chemist obtains from "dead albuminoids", by hydrolysis, amino acids and amides, the organisms of birds or snakes transform the "living albuminoids" into the "cyanogen compound" uric acid. Pflüger points to the presence of the radical cyan in a number of products of animal metabolism: creatine, creatinine, guanine, hypoxanthine, xanthine, etc. It is also the case with the sulfocyanate of saliva. None of these products, Pflüger states, can be obtained from "dead albuminoid".

Against these inductions, Pflüger had of course to face the fact that urea, an amide, was known as the main nitrogenous excretion product of mammals. To this he retorts that no chemical treatment of albuminoid has led to urea, while this compound can be obtained from uric acid, creatine, creatinine or ammonium cyanate, which in his understanding represents the ultimate wheel-work of the watch taken to pieces. ("Das cyansäure Ammonium repräsentiert uns ein Stück Lebensprozess, den Letzten Ablauf der aufgezogenen Uhr [3] . . ." p. 335.)

Finally, Pflüger considers the urea excreted by mammals as a derivative of uric acid.

3. Intellectual sources of Pflüger's biochemical philosophy

Besides the concept of intracellular respiration, which became a paradigm of physiology, Pflüger's theory attempted to provide a molecular basis for this cellular respiration. He conceived the idea of the protoplasmic albuminoids as huge polymers of variable size, which he describes as labile energy-rich compounds characterized by the presence of cyanogen radicals.

Pflüger acknowledges his indebtedness to Liebig's theory of the source of muscular energy in proteins (see Chapter 7). Pflüger was also, as mentioned above, influenced by the concept of "inogen substances" of muscle (Kühne, Hermann) [4]. But one of the most interesting aspects of Pflüger's theory is its interplay with the bioenergetics of the period.

It has been a constant tendency to identify the sources of energy in biosynthesis with what had previously been recognized as a source of mechanical work. This applies to the derivation of the idea of the formation of high-energy proteins from the "inogen" concept at the level of muscle, as well as to the recognition of the source of energy of biosynthesis in high-energy bonds of ATP, first recognized at the level of muscle energetics (see Chapter 23). But we should keep in mind that the long period of dominance of the protoplasmic theory coincides with the time of the development of the basic concepts of bioenergetics. Huxley formulated the "protoplasm theory" in 1868. The lecture of Helmholtz, *Ueber die Erhaltung der Kraft* was published in 1847 (privately, after having been refused by the editor Poggendorff). It was only by the end of the 19th century that the Newtonian concepts of space, time, mass and *force* were replaced by those of space, time, mass and *energy*. In mechanics, what appeared as conserved was the sum of potential energy (Spannkraft) and kinetic energy (*vis viva*, lebendige Kraft). It was to the latter that the term energy was first applied by Young in 1807 (see Elkana [5]).

As remarked by Fruton [2],

"Pflüger believed that he was applying to the problem of energy transformation in biological systems the most modern ideas of physics, as expressed in Clausius's formulation of the second law of thermodynamics" (p. 18).

"The emergence of the protoplasmic theory of life was paralleled, during 1850—1870, by the development of molecular theory along two major tracks. One of those was the one leading from the study of the diffusion of gases and of substances in solution. The phenomena that were observed were explained by a theory that defined a molecule as the smallest portion of a substance that moves as a whole (Maxwell [6]). Among the triumphs of this theory was the mathematical description of heat as a mode of molecular motion. Thus, for Pflüger it was a matter of some importance that Clausius had shown that the difference between what Clausius called the "living energy" associated with the molecular motion of a system and the heat content of that system is greater as the number of atoms composing a molecule increases (Clausius, p. 309) . . ." (p. 19).

The other track considered molecules in chemical reactions, defining the molecule

"as the smallest particle of a substance that retains the properties of that substance in reactions that conform to the laws of definite and multiple proportions. There was emerging, as a counterpart to the mathematical theory of molecular motion, a geometrical theory of the arrangements of atoms in a molecule, partly based on the recognition that some reactions involved the rearrangements of such atoms was given to intramolecular motion involving atoms held together by valence forces. From the reactivity of particular atomic groups, and the amount of heat evolved in their reactions, inferences were drawn about what was called the *Spannkraft* of these groups; this term was intended to connote something akin to the potential chemical energy associated with the interatomic vibrations" (p. 20).

As noted by Fruton [2] (p. 18), Pflüger was also applying the most modern concepts of organic chemistry as developed in the treatise of Kekulé, and the most modern work on protein as represented by the work of Hlasiwetz and Haberman, and of Ritthausen (see Chapter 45).

Proposed by a leading German physiologist, the successor of Helmholtz at Bonn, Pflüger's theory won very wide recognition during the remaining part of the 19th century. It appeared in a very long and elaborate paper which is justly recognized as a classic, in having introduced one of the basic concepts of modern biochemistry, intracellular respiration, as we have shown in Chapter 8 (pp. 164—170). One of the attractions of Pflüger's theory was its modernism, its endeavor to reduce physiological phenomena to a "molecular" basis. Pflüger related to the most exciting aspects of the new thermodynamics the conceptual scheme he proposed: the conversion of dead proteins into living proteins, i.e. labile energy-rich structures formed by polymerizations resulting in a change in which nitrogen atoms combine with carbon atoms to form cyanogen-like relations. At death, the living proteins returned to the stable form known by the chemist.

"Living protein" contained potential energy (Spannkraft) and was able to liberate "living force" or "lebendige Kraft". This should not be confused with "vital force", but recognized as synonymous with kinetic energy. On the other hand, Pflüger's "living proteins", as well as Huxley's "protoplasm", belong to a tendency which Kohler [7] has aptly characterized as

"the striving of biochemists for a comprehensive explanatory concept on the biochemical level, between the passive metabolite and the organized cell" (p. 292).

4. Variations on Pflüger's theme

As remarked by Fruton [2]:

"For at least 40 years afterwards, there was a succession of variations on this theme" (p. 20).

Different structures were considered as responsible for the lability of living albuminoids, but a common feature of these theories was that

"the chemical instability of cellular constituents was directly related to the energy that could be derived from its decomposition. Even after attention had shifted during the present century from "living" protein molecules to low-molecular weight substances such as creatine phosphate, this connection between chemical instability and the amount of energy released was a recurrent note in biochemical thought" (Fruton [2], p. 21).

The cyanogen theory of Pflüger was, as noted by Fruton [2], again referred to when adenine was recognized as a constituent of "nuclein". At the time, the site of biosynthesis was considered as being the cell nucleus (see Coleman [8]).

To quote Chittenden [9]:

"Such being the nature of adenine, it is not to be doubted that bodies emanating from this substance with strong affinities must be important factors in the physiological and chemical processes, especially those of a synthetic order, going on in all cellular tissues. In this connection it is to be remembered that Pflüger, on purely theoretical grounds, ascribed great importance to the physiological role played by the cyanogen group with polymerization, etc., in the living albumin molecule. . . . In the discovery of adenine and its close relationship to the typical xanthine bodies we have added proof of the existence of cyanogen-containing radicals in the protoplasm of the cell, especially in the karyoplasm of the nucleus. In all of these xanthine bodies there is to be seen a peculiar combination of "carbon, nitrogen and hydrogen such as not found in dead protein matter" (pp. 115—116, cited after Fruton [2] p. 22).

The idea of a difference between the chemical composition of dead and of living matter has returned ever and anon. For instance, we read in a paper of Nencki [10] published in 1885:

"As Pflüger stated rightly over ten years ago, the protein of the living cell must have an entirely different molecular structure from that of dead tissues" (p. 343 translated by Fruton [2], p. 25).

The distinction between living proteins and dead proteins appears also in the classical book of Ehrlich on cellular oxidations. As we have stated in Chapter 27 (p. 205) Ehrlich developed the idea that the living proteins did attract "side chains" which became the effectors of biochemical changes.

Descending from Pflüger's philosophy came the concept, introduced by Verworn, of "biogen", a huge "molecule" endowed with the properties of Pflüger's energy-rich albuminoids, as well as with the side chains of Ehrlich (see Introduction to Part III, p. 7 and Chapter 17, p. 23).

5. Loew's theory of the polycondensation of aminoaldehydes into energy-rich proteins

Among the variants of the theories of energy-rich proteins, Loew's theory [11] deserves a special mention. Loew did not start from food proteins as precursors but from smaller units, the aminoaldehydes.

He considered the aldehyde group, the chemical activity of which was increasingly recognized around 1870, as responsible for the potential energy of "living proteins". Loew proposed that the biosynthesis of proteins in protoplasm is accomplished by a polycondensation of aminoaldehydes, in particular of the dialdehydes of aspartic acid. Loew remains known for his identification of catalase (see Chapter 26, p. 195). He was born in 1844 and died at the ripe age of 96 years during World War II. He had worked in Liebig's laboratory before he left Germany in 1867 for the United States where he occupied several positions during the following decade.

Returning to Munich he became a member of Nägeli's staff at the Institute of Plant Physiology, which he left in 1892 to become professor of agriculture in Tokyo, where he remained until 1907, except for a stay (1896–1900) at the U.S. Department of Agriculture. From 1907–1913 he assumed the direction of the section on plant physiology of an agricultural Institute in Puerto Rico. In 1913 he came back to Munich as head of the biochemical division of the Institute of Botany. In 1923 he went to Brazil, to return in 1926 to Berlin, where a factory manufacturing a product developed by him insured his subsistence until his death in 1941 (on Loew, see Klinkowski [12] and Fruton [2]).

The background of the aldehyde theory was the presence of aldehyde groups in "protoplasm" as claimed by Loew and Bokorny [13] in a book which intends nothing less than to define the chemical causes of life. The authors claim that the characteristic feature distinguishing living plant proteins from dead plant proteins is the presence of aldehyde groups in the former. The evidence is entirely based on the reduction of a dilute solution of alkaline silver nitrate. The test is, the authors claim, positive with living plant proteins and negative with dead proteins. According to Loew, the "active protein" of protoplasm is a large molecule for which he adopts the empirical formula $C_{72}H_{112}N_{18}O_{22}S$ (which had

References p. 131

Plate 144. Oskar Loew.

been proposed by Lieberkühn in 1851), and which according to his estimation contained 12 aldehyde groups, the instability of which was the true cause of life. Death was a rearrangement of aldehyde groups in various ways. The theme survived for a long time and is found, for instance, supported by Nencki in 1900 (Bickel [14]). Loew continued to defend it into the 1920's and more recently it reappeared in a paper by Delbrück [15] published during the last year of Loew's long life.

6. The protoplasmic gel

One of the forms taken by the protoplasm theory was its variety oriented towards biocolloidology, introduced by Graham [16] in 1861, and according to which life phenomena are the result of the colloidal state of matter, opposed to the crystalloid state. We shall, in Chapter 44 record premonitory aspects of the idea in Schwann's cellular theory. Typical of this position is the statement of Graham, including the recourse to capital letters:

"The colloids possess ENERGIA. It may be looked upon as the probable primary source of the force appearing in the phenomena of vitality" [16].

Seventy-five years later, we find a similar profession of faith by Gortner [17]:

"All of the reactions and interactions which we call life take place in a colloidal system, and the author believes that much of the "vital energy" can in the last analysis be traced back to energies characteristic of surface films and interfaces" (see Chapter 14, p. 280).

In spite of the similarity of Pflüger's "mass molecules" and the "micelles" of the biocolloidologists, the difference between the biocolloidal protoplasmic theory and Pflüger's theory, on similar views, is that in the latter, the source of energy was considered as a high-energy bond and not surface energy.

According to biocolloidal theories, biological phenomena were the results of changes introduced by inorganic ions in processes of agglutination, lysis, dispersion, hydration and dehydration of colloid micelles, believed to compose the protoplasmic gel. The biocolloidal theory states that the source of energy in organisms can be traced back to surface film energy and interface surface energy, the structure and size of the micelles changing continuously due to

References p. 131

Plate 145. Thomas Graham.

association and dissociation phenomena at the level of all kinds of micellar aggregation states.

7. Bernard's "vital creation"

We have in Chapter 10 (pp. 206—207) formulated the nature of Bernard's view on metabolism and emphasized the penetrating nature of his views on indirect nutrition as well as his recognition of what he called "vital creation" as a hereditary structure. In order to go deeper into the concept of "vital creation", we may peruse the last publication of Bernard (1878), his *Lectures on the Phenomena of Life Common to Animals and Plants*, vol. I [18]. Bernard's adherence to Huxley's protoplasm theory is clearly recognized and he states that protoplasm is the organic basis of life. When he has related the ideas formulated by the chemists of his time to account for the most fundamental of the biosynthetic phenomena, photosynthesis, Bernard adds:

"These concepts are strongly impregnated with what can be called *Synthetic chemistry*. . . . *Natural chemistry* is perhaps entirely different; it would be possible for example that all the syntheses invented by the chemists are without reality, and that the immediate principles are produced by way of decomposition and cleavage of a single and identical material, the protoplasm" (translation by Hoff et al. [18], p. 155).

Bernard had already formulated this view in an entry in his notebook known as the *Cahier rouge* as early as December 1856:

"The vital creation of tissues is neither chemical nor physical.

The disaggregation and decomposition of tissues is physical and chemical.

Between these two orders of phenomena there is a strict line of demarcation" (translation by Hoff et al. [19], p. 35).

Chronologically this entry in the *Cahier rouge* corresponds with the brilliant discovery of the glycogenic function of the liver. Referring to this discovery in the *Lectures on the Phenomena of Life* Bernard [18] writes:

"Thus, in animals as well as in plants, sugar is formed at the expense of the starch. The formation of this starch in these two kingdoms is considered as an act of organic creation, a synthesis. The formation of sugar is on the contrary an organic destruction, a hydration of the starch which brings about its conversion into dextrine and into glucose. . . ." (p. 165).

The view of the essential difference between chemical decomposition and vital creation is based on Bernard's experiments concerning the perfused liver, in which, in an isolated liver, the sugar-like material is washed away by perfusion. After the organ is set aside for a time, a new quantity of sugar is formed. The test can be repeated several times until exhaustion of the glycogen. From this, Bernard concluded that in this dead organ the chemical process (decomposition) continues while "vital creation" ceases. In his mind, the glycogen, the product of vital creation,

"appears, not by a true synthesis in the chemical sense of the word, but by a cleavage of the protoplasmic material" (p. 172).

8. Retrospect

Considering the aspects of the varieties of protoplasm theory in relation with the origin of energy in cells, we have stressed in the *Introduction* to Part III the retarding effect it has exerted upon the chemical studies which led to the recognition of the high-energy bond of ATP, identified as the universal energy source in metabolism. As we have noted, the reaction against the irrationalist conceps of protoplasmists was vigorously led by Hopkins.

The reversed pathway, the formation of mass molecules or of micelles, huge polymers of variable size, the variability of which was considered as the basis of a number of activities, has acted as an epistemological obstacle to one of the factual acquisitions which enriched the conceptual system of modern biochemistry: the recognition of the proteins as truly defined macromolecules (Chapter 15). We have, in Chapter 14, insisted on the retarding influence of biocolloidology on the progress of biochemical knowledge. This judgement does not concern the aspects of colloidal chemistry which have been useful to chemists, pharmacists, etc. but regards the theory based on micelle surfaces, applied in an irrelevant way to biological systems. At the time of their formulation, the theories concerned with giant protein molecules (mass molecules, micelles) exerted an influence in the field of ideas on immunology and pharmacology. This point has been made by Parascondola in the course of a discussion on energy-rich proteins, after the presentation of the paper by Fruton [2] referred to above.

To quote Parascandola [2], who takes P. Ehrlich as an example of a scientist whose conceptions arose from the theories of giant molecules,

> "Ehrlich adopts the idea that the cell is a protoplasmic molecule with essentially nutritive side chains, oxidative side chains, etc. This is the same concept which he later applies to immunology to explain the formation of antibodies and eventually it is the same concept that he applied to the receptors for drugs, the so-called chemoreceptors. And Loew also uses this kind of concept when he is talking about poisons; he suggests that many poisons act by tying up these aldehyde or amino groups. It is interesting that Ehrlich opposed this view even though he supported Loew's other work. He opposed the view that chemicals and poisons acted in this way mainly because one can extract most chemicals and poisons from the body like alcohol. So he very clearly distinguished for many years between the toxins which are assimilated into the giant protein molecule and chemical poisons like alkaloids which are not and which tend to act through a kind of physicochemical process like a solid solution in lipids. Eventually, after the work of Langley on receptors, he accepted this idea of receptors and came to apply them to the same concept, where drugs are essentially assimilated by actually combining with a side chain of the cell. This theory was influential in pharmacology at least. It was only actually in the 1920's that perhaps the influence of this began to be less felt and people began to search for more specific receptor enzymes, things like glutathione — rather than these side chains of protoplasm" [2] (p. 43).

Parascandola [2] remarks that much more criticism of the biogen hypothesis began to appear in the period 1900—1920 and he refers to Bayliss calculating that if strophantin combined with the giant molecule of muscle cell, this molecule should have a molecular weight of over 57 millions, which seemed impossible. The progress of the enzyme theory (see Chapter 13) displaced the attention from biogens to enzymes recognized as definite molecules, shifting the situation of "receptors" to smaller units.

The influence of the giant molecule or micelle concept in the field of ideas during the second half of the 19th century belongs to what we have called horizontal history and as such deserves the interest of historians. The influence of biocolloidology lasted longer than that of the mass molecules of Pflüger. The rejection of the concept of the micellar nature of proteins was described in Chapter 15. It was based on the work of E. Cohn and of his school on proteins as polyelectrolytes. As noted by Fruton [20] (p. 143), the theoretical basis of this advance, one of the milestones in the transition toward modern biochemistry, is to be found in two advances published in 1923. One is the theory of Debye and Hückel

Plate 146. Paul Ehrlich.

dealing with the behavior of an ion in solution, in relation to the charge and radius of the ion, to the other ions in solution and to the dielectric constant of the solvent. The second insight was the recognition by Bjerrum that the amino acids in their isoelectric state were dipolar ions, i.e. charged molecules. In the isoelectric state, the proteins, in spite of having a net charge equal to zero, were highly charged.

The factual basis of Pflüger's theory of energy-rich "mass molecules" was slim. It was essentially based on the exergonic character of the decomposition of cyan polymers, and the excretion of uric acid by birds and snakes. The irrelevant character of such factual knowledge in relation to the conceptual aspect derived from it became obvious and as a consequence the concept was dismissed from the conceptual system of biochemistry. The factual basis of Loew's theory was the assumed reduction of silver nitrate by "living proteins". Not only was this irrelevant, but it was known that a variety of substances can reduce silver nitrate (Baumann [21]).

The aspects dealt with in the present chapter depend, from the point of view of the history of ideas, on the protoplasm theory, and on its basic (and irrational) contention of vital creation by which dead food materials are transformed into living protoplasm. It is this concept which we find again, in the last publication of Bernard [18], in 1878, substantiated with new factual evidence. As stated above, this factual evidence in favor of vital creation turned out to be recognized later as a post-mortem hydrolysis of glycogen. This interpretation was not available at the time but, no doubt, the one given by Bernard was a symptom of his metaphysical stance.

The brilliant discovery of the glycogenic function of the liver was a striking demonstration of the need for experimental procedures applied in situ, at the seat of metabolic processes, and it remained a fertile teaching in the field of biochemical methodology. It is sometimes stated that we owe to Bernard the establishment of an adequate interplay between chemistry and physiology which would insure the development of modern biochemistry.

It is also sometimes claimed that the concept of a biosynthetic pathway was introduced by the glycogen synthesis discovered by Bernard. This discovery was indeed a milestone and it did, in line with the polemics which had been raging for decades concerning

References p. 131

the synthetic potentialities of animal organisms, convincingly strengthen the acceptance of such potentialities. This aspect is expressed in the following quotation:

"Amid the speculations, animal synthesis was finally demonstrated with Claude Bernard's momentous discovery of hepatic glycogenesis. This destroyed the foundation of Dumas' vital dualism and provided the experimental method which all previous attempts had failed to find. The way in which Bernard combined chemistry with physiology constituted the beginnings of a true biochemistry" (Goodman [22], p. 122).

While paying a deserved tribute of admiration to the importance of the discovery of hepatic glycogenesis by Bernard, this statement requires definition. First of all, when Bernard discovered hepatic glycogenesis, a demonstration of the biosynthetic abilities of animals, the "chemists" had already given up the dualist theory. It was, as stated above, Boussingault, who provided the final proof of fat biosynthesis by animals. Vital dualism was rejected. On the other hand, the brilliant work of Bernard did focus attention on biochemical events considered in situ and on the biochemical experimentation on organisms. But, by attributing the biosynthesis of glycogen to a consequence of protoplasm biosynthesis due to a "vital creation" of a non-chemical nature, he abandoned himself to irrational tendencies which, to say the least, led off on a sterile track.

On the other hand, the "chemists" did, during the polemics on fat biosynthesis, change their views, as was stated in Chapter 42, and Liebig did also.

To quote Holmes [23]:

"Had the followers of their views not been able to sift out the fruitful conceptions and methods from the fanciful excesses, the approach they represented could not have had the lasting influence it actually attained" (p. 448).

In fact, Dumas, Liebig and their followers, after giving up their imaginary views, deprived of experimental basis, opened the way which ultimately led to the unravelling of biosynthetic pathways. The chemists were not contaminated by the protoplasmic myth which led Pflüger and Bernard astray. Though a purgatory of hypothetical views had to be crossed, methodological weapons were finally available for the chemical unravelling of biosynthesis, as will be shown in Part V of this *History*.

REFERENCES

1 D. Diderot, Le Rêve d'Alembert, in Oeuvres Complètes, vol. 2, Paris 1875.
2 J.S. Fruton, in Proceedings of the Conference on the Historical Development of Bioenergetics, sponsored by the American Academy of Arts and Sciences, October 11—13, 1973, Boston, 1975, pp. 17—35.
3 E. Pflüger, Arch. Ges. Physiol., 10 (1875) 251, 641.
4 D.M. Needham, Machina Carnis. The Biochemistry of Muscular Contraction and its Historical Development, Cambridge, 1971.
5 Y. Elkana, The Discovery of the Conservation of Energy, London, 1974.
6 J.C. Maxwell, Nature, 8 (1973) 437.
7 R.E. Kohler, J. Hist. Biol., 8 (1975) 275.
8 W. Coleman, Proc. Am. Phil. Soc., 109 (1965) 124.
9 R.H. Chittenden, Am. Naturalist, 28 (1894) 97.
10 M. Nencki, Arch. Exp. Pathol. Pharmakol., 20 (1885) 332.
11 O. Loew, Arch. Ges. Physiol., 22 (1880) 503.
12 M. Klinkowski, Ber. Dtsch. Chem. Ges., 74 A (1941) 115.
13 O. Loew and T. Bokorny, Die chemische Ursache des Lebens theoretisch und experimentell nachgewiesen, Munich, 1881.
14 M.H. Bickel, Marceli Nencki 1847—1901, Bern, 1972.
15 M. Delbrück, Cold Spring Harbor Symp., 9 (1941) 122.
16 T. Graham, Phil. Trans. Roy. Soc. (London), (1861) 183.
17 R.A. Gortner, Outlines of Biochemistry, 2nd ed., New York, 1838.
18 C. Bernard, Lectures on the Phenomena of Life Common to Animals and Plants, vol. I, Springfield, Ill., 1974 (translation by H.E. Hoff, R. Guillemin and L. Guillemin, from Cours de Physiologie générale du Muséum d'Historie Naturelle. Leçons sur les Phénomènes de la Vie Communs aux Animaux et aux Végétaux, Paris, 1878).
19 C. Bernard, Le Cahier Rouge (translated by H.E. Hoff, L. Guillemin and R. Guillemin), Cambridge, Mass., 1967.
20 J.S. Fruton, Molecules and Life. Historical Essays on the Interplay of Chemistry and Biology, New York, 1972.
21 E. Baumann, Arch. Ges. Physiol., 29 (1882) 400.
22 D.C. Goodman, Med. Hist., 16 (1972) 113.
23 F.L. Holmes, Claude Bernard and Animal Chemistry. The Emergence of a Scientist, Cambridge Mass., 1974.

Chapter 44

Crystallization and Biosynthesis

1. Crystallization as an overgeneralization

Biologists have always been fascinated by the growing arborescences observed in the crystallization of different inorganic substances. Such arborescent growths had been classified by chemists in a mythological guise as arbre de Diane or silver tree (silver amalgam), arbre de Mars (iron silicate and potassium carbonate), arbre de Jupiter (tin precipitated by zinc), arbre de Saturne (crystallization formed around a zinc plate dipped in lead acetate), etc. The forces responsible for such arborizations have repeatedly been compared to those active in the formation of living organisms. Along those lines, references are found to crystallization in the writings of F. Bacon, Hooke and Maupertuis (see Hall [1], *passim*).

The comparison of the formation of whole organisms to crystallization belongs to that category of concepts which have been characterized by Bachelard [5] as bad themes of generality based on rapid and superficial syntheses.

In her book on crystallography, Hélène Metzger [2] identified a number of such comparisons in the writings of the 18th century and for instance in the work of Buffon, of Guéneau de Montbéliard, who presents spontaneous generation as a crystallization (Metzger [2], p. 112) and of de la Métherie [2,3]. The latter exposed his concept of the formation of organisms by crystallization in many writings, and among others in his book entitled *Vues Physiologiques sur l'Organisation Animale et Végétale* [4]. Doctor of medicine and editor of the *Journal de Physique, de Chimie et d'Histoire Naturelle et des Arts* (which ceased publication in 1823 and in which, in 1811, the classical paper of Avogadro was pub-

References p. 143

lished), de la Métherie was an ardent phlogistonist and he acted as an obstinate opponent of Lavoisier. His biological philosophy can be considered as belonging to the theory of fibres.

Concerning biosynthetic processes, which he considers at the level of tissues, de la Métherie [4] exposes his theory as follows:

"How is the calcareous matter deposited in the bones, layer by layer, as M. Duhamel did prove through his madder-wort experiments? How is the gelatinous part deposited in muscle and viscera, and the fatty part in cellular tissue? Such is the mechanism of nutrition.

"As we have said, we believe that it is by affinity that the calcareous part gets deposited in the bones, the gelatinous part in muscles, the bile in the liver, etc.; but the ultimate cause governing the deposition of such parts at the sites designated by their nature, appears to me as being the force responsible for the universal crystallization of matter. In a phial containing several salts, each crystallizes independently. In the vessels of animals and plants, several liquors are mixed and each is deposited in a particular place. This is all that we know. Each part of matter has received its own force responsible for its crystallization governed by its configuration. Metals crystallize, stones crystallize, salts crystallize, gums, resins, extracts, resinous extracts, gummy materials crystallize. The organic fluids of animals and plants must also crystallize and they do it under more or less agreeable forms, such as silver does in a silver tree" (pp. 79—80) (translation by author).

De la Métherie considers the phenomena of animal generation as a kind of crystallization.

"The seeds of the male and of the female, being mixed, act as two salts would and the result is the crystallization of the fetus" (translation by author).

Everything crystallizes in nature, even the celestial bodies. In this respect, de la Métherie was expressing the kind of overgeneralization that was relied upon earlier when Wallerius considered coagulation as explaining the formation of all aspects of the Universe (see Bachelard [5], p. 63).

"Several great physicists, under the pressure of factual evidence, do not hesitate to recognize the existence of spontaneous generations. They do not hesitate to state that molds, microscopical animals (which exist in vast numbers), the different helminths found in human bodies, the flukes of sheep's liver, etc. are not produced, as is the case for other animals by a father and a mother, and that they result from the adoption of new forms by animal parts. These animal parts have only been able to assume the new form by crystallizing in such or such manner, as an expression of their forces and configurations" (p. 286) (translation by author).

There is a great similarity between the concepts of de la Métherie and those of Reil [6]. Considering the "fibres" as impenetrable solid bodies, he believes that they attract nutrients which are added externally in a process of "tierische Kristallisation".

2. Cell formation compared to a crystallization

The cell theory has two aspects, as pointed out by Virchow (Chapter 6, p. 136), One of these aspects, introduced by Schleiden and accepted by Schwann, was the erroneous concept of the formation of the cytoblast (nucleus) around a nucleolus arising in a cytoblastema. The other aspect, which is due to Schwann and which has enjoyed a permanently increasing acceptance, is the recognition of all organisms as derived from nucleated cells, differentiated into various kinds of "elementary particles" (in the language of Schwann).

The erroneous theory of the formation of nuclei, adopted by Schwann [7], is formulated by him as follows:

"A nucleolus is first formed; around this a stratum of substance is deposited, which is usually minutely granulous, but not as yet sharply defined on the outside. As new molecules are constantly being deposited in this stratum between those already present, and as this takes place within a precise distance of the nucleus only, the stratum becomes defined externally, and a cell nucleus having a more or less sharp contour is formed" (translation by H. Smith [7], p. 175).

The background of this theory is to be found in the ideas of Schleiden regarding the formation of plant cells (Uhrglastheorie, see Chapter 6, p. 136), and in the observations of Wagner. To quote Schwann:

"In plants, according to Schleiden, the nucleolus is first formed, and the nucleus around it. The same appears to be the case in animals. According to the observations of R. Wagner on the development of ova in the ovary of *Agrion virgo*, the germinal spot is first formed and around that the germinal vesicle, which is the nucleus of the ovum-cell, Eizelle" [7] (pp. 174—175).

When the nucleus is formed, the formation of cells, according to Schwann, takes place as follows:

"A stratum of substance, which differs from the cytoblastema, is deposited upon the exterior of the nucleus" (p. 176).

As we have recalled in Chapter 6, the last section of the third part of the *Mikroskopische Untersuchungen* [7] is entitled "Theory of the cells" (pp. 186—215), and is devoted to theoretical aspects. It is in this theoretical section that Schwann draws a comparison between cell formation (as described in the cell theory) and crystallization. The analysis of the formation of the cell substance by accretion of molecules (and as we have stated, Schwann considers the chemical molecule) and the comparison of this accretion with that of molecules in crystals has sometimes been considered as an extreme expression of Schwann's physicalism.

There is nothing, in Schwann's analysis, comparable to the sweeping statements found, for instance, in the writings of de la Métherie. On the contrary, Schwann performs a painstaking examination of the concepts involved and he is careful to point to the differences between both phenomena. How cautiously Schwann approaches the comparison appears from the following quotation:

"And now, in order to comprehend distinctly in what the peculiarity of the formative process of a cell, and therefore in what the peculiarity of the essential phenomenon in the formation of organized bodies consists, we will compare this process with a phenomenon of inorganic nature as nearly as possible similar to it. Disregarding all that is specially peculiar to the formation of cells, in order to find a more general definition in which it may be included with a process occurring in inorganic nature, we may view it as a process in which a solid body of definite and regular shape is formed in a fluid at the expense of a substance held in solution by that fluid. The process of crystallization in inorganic nature comes also within this definition and is, therefore, the nearest analogue to the formation of cells.

"Let us now compare the two processes, that the difference of the organic process may be clearly manifest. First, with reference to the plastic phenomena, the forms of cells and crystals are very different. The primary forms of crystals are simple, always angular, and bounded by plane surfaces; they are regular or at least symmetrical, and even the very varied secondary forms of crystals are almost, without exception, bounded by plane surfaces. But manifold as is the form of cells, they have very little resemblance of crystals; round surfaces predominate, and where angles occur, they are never quite sharp, and the polyhedral crystal-like form of many cells results only from mechanical causes. The structure too of cells and of crystals is different. Crystals are solid bodies, composed merely of layers placed one upon another; cells are hollow vesicles, either single, or within another" [7] (pp. 200—201).

The last sentence is one of the sources of difficulty in the interpretation of Schwann's writings, particularly with respect to semantic aspects. Such structures as a muscle fibre or a nerve fibre he calls "elementary particles", but he considers them as "modified cells" resulting from differentiations of embryonic cells similar to plant cells. He sees these cells in the hollow form but he soon recognizes other forms and he defines the cell, whatever its characters may be and whether embryonic or differentiated into "elementary particle", as a succession of layers around a nucleus. There is no doubt that he recognizes such an "elementary particle" as a muscle fibre as a modified solid cell. The simile which Schwann draws with respect to crystals and cells results in a model formation in which he does figure out how bodies capable of imbibition would behave if they could crystallize. Preoccupied with finding an interpretation of the accretion of molecules in the successive layers surrounding the nucleus in cells and in "elementary particles" (differentiated cells), Schwann, probably influenced by the concept proposed in 1837 by Laurent of a formation of organic molecules by successive layers, as in crystals (see Chapter 12, p. 255), attempts to understand the formation of the "cell layer" (Zellenschichte) around the nucleus as a deposition of organic molecules. This accretion and the attraction of new molecules he considers as the result of the existence of phenomena similar to those which govern the accretion of molecules in a crystal. In the case of cells, the soft material, capable of imbibition, is also deposited in layers. Schwann remarks that, against a concept of the action in cell formation of attracting intermolecular forces analogous to those governing the formation of crystals, objections may be formulated on the basis of the selective nature of the cellular accretion.

"The attractive power of the cells manifests a certain degree of election in its operation; it does not attract every substance present in the cytoblastema, but only particular ones; and here a muscle cell, there a fat cell is generated from the same fluid, the blood. Yet crystals afford us an example of a precisely similar phenomenon, and one which has already been frequently adduced as analogous to assimilation. If a crystal of nitre be placed in a solution of nitre and sulphate of soda, only the nitre crystallizes; when a crystal of sulphate of soda is put in, only the sulphate of soda crystallizes. Here, therefore, there occurs just the same selection of the substances to be attracted" [7] (p. 211).

But Schwann acknowledges that the mere recognition of the common feature of a deposition of a solid substance from a fluid does not suffice to assume any more intimate connection between crystal formation and cell formation by the deposition of layers around a nucleus.

"But we have seen, first, that the laws which regulate the deposition of the molecules forming the elementary particles of organisms are the same for all elementary parts; that there is a common principle in the developments of all elementary parts, namely, that of the formation of cells; it was then shown that the power which induced the attachment of the new molecules did not reside in the entire organism, but in the separate elementary particles (this we call the plastic power of the cells); lastly it was shown that the laws, according to which the new molecules combine to form cells, are (so far as our incomplete knowledge of the laws of crystallization admits of our anticipating their probability) the same as those by which substances capable of imbibition would crystallize. Now the cells do, in fact consist only of material capable of imbibition; should we not then be justified in putting forth the proposition, that the formation of the elementary parts of organisms is nothing but a crystallization of substances capable of imbibition, and the organism nothing but an aggregate of such crystals capable of imbibition?" [7] (p. 212).

As was noted by Maulitz [8], Schwann insists on the hypothetical characters which he is careful to confer to the theory, based on analogy.

The scope of Schwann, in this section of the *Mikroskopische Untersuchungen*, is to enquire into the mechanism of accretion of new molecules in the cell layer. He very carefully draws a comparison with the attraction of molecules in crystals, to emphasize the participation of intermolecular attraction in the process of the self-assembly of intracellular components. But he is conscious of the difference between crystals and cells resulting from the high water content of the latter: this he designates as their faculty of imbibition.

"Most organic bodies are capable of being infiltrated by water, and in such a manner that it penetrates not so much in the interspaces between the elementary tissues of the body, as into the simple structureless tissues, such as areolar tissues, etc.; so that they form an homogeneous mixture, and we can neither distinguish particles of organic matter, nor interspaces filled with water. The water occupies the infiltrated organic substances, just as it is present in a solution, and there is as much difference between the capacity for imbibition and capillary permeation, as there is between a solution and the phenomena of capillary permeation. When water soaks through a layer of glue, we do not imagine it to pass through pores, in the common sense of the term, and this is just the condition of all substances capable of imbibition. They possess, therefore, a double nature, they have a definite form like solid bodies; but like fluids, on the other hand, they are also permeable by anything held in solution" [7] (p. 203).

When he was appointed professor of anatomy at the Catholic University of Louvain, where he arrived in March 1839, Schwann began his course of lectures on April 23. He delivered the lectures on general anatomy between July 16 and 30, 1839. In the academic year 1840—1841 this section of lectures devoted to general anatomy took place between July 22 and August 4, 1840. During the next academic year, 1841—1842, Schwann started his course with the lectures on general anatomy (October 25, 1841 to January 8, 1842) for which he had, during the previous summer, prepared a written text and from which he continued to dictate during the following years, adding from time to time alterations or additional material. In the first lecture of the manuscript text, dated 1841, Schwann [9], commenting on the imbibition state of the cytoplasmic material, suggests that this represents a fourth aggregation state, that should be added to the gaseous, liquid and solid states.

"The best microscopes do not permit to distinguish spaces occupied by water from other spaces filled with organic substances. And nevertheless water is not there in a chemical combination: we may increase or decrease at will the quantity of water present in a given substance. . . In order to get an idea of the nature of this state of imbibition we may consider a solution of glue. As long as it remains in the liquid state, water persists in it as in other solutions and the molecules of water remain intimately mixed with those of the dissolved material. But if the solution of glue coagulates and becomes solid, the molecules of water remain as intimately dispersed between the molecules of glue as in the solution, though the substance has become solid" (translation by author).

It is clear that when he identifies, as he does in the *Mikroskopische Untersuchungen*, the substance of the organisms as "the form under which substances capable of imbibition crystallize", he foreshadows the concept of the gel of colloid chemists. Schwann himself confirmed this interpretation by adding to the text cited above in his lecture notes, twenty years later, the following lines:

"Graham divides up liquids between crystalloids (sugar, salt, etc.) and colloids (glue, albumin, gum, caramel, extractive subtances). The first ones have an affinity for water and raise the ebullition point: the colloids do not. Separation by dialysis through parchment" (translation by author).

3. The "micellae" of Nägeli

Among the different hypothetical theories concerned with "metastructural" particles of protoplasm as conceived according to

References p. 143

Plate 147. Carl Wilhelm von Nägeli.

Huxley, we have mentioned the "micellae" of Nägeli (Chapter 16, p. 296). We return to this theory here in consideration of its crystallographic implications, referring the reader, for a closer analysis, to Chapter 49 of Hall's book on *Ideas of Life and Matter* [1]. The background of Nägeli's theories is to be found in his studies on starch granules of plant cells. From their swelling in water and their birefringence he reached a generalized concept of organized materials composed of water-encased "protein crystalloids". These he considered as crystalline birefringent "molecules" associated in a defined arrangement. We find in the conceptual system of Nägeli the idea of an arrangement similar to that of a crystal and the idea of imbibed structures, as in Schwnn. But while the molecule of Schwann is the molecule of the chemist, the "molecule" of Nägeli is itself similar to a crystal and is called by him a "micella". He considers the micellae as assembled in meshes.

Nägeli's theories were applied by him to a host of biological concepts, but they were among the theoretical views which, along with the concepts of biocolloidology, were rejected when adequate methods allowed the recognition of the existence, in cytoplasm, of definite organelles, an advance which was discussed in Chapter 16 and which introduced one of the main features of the conceptual system of present-day biochemistry.

4. Bernard's comparisons between mineral forms and living forms

In the first lecture of his last publication, the *Lectures on the Phenomena of Life Common to Animals and Plants* [10] (1878), Bernard, with respect to the tendency of organisms to prove their individuality by re-establishing their form, refers to the statement by Tiedemann according to which this tendency appears as an exclusive character of life:

"This is not correct; crystals, like living bodies, have their form and their particular plan, and when disturbing actions within the ambient milieu displace these from them, they are capable of re-establishing them by a true healing or reintegration of the crystal".

Pasteur has seen

"that when a crystal has been broken in any one of its parts, and when it is replaced in its mother liquor, it is seen that at the same time that the crystal increases in all its dimensions by a deposition of crystalline particles, a very active work takes place on the broken or deformed parts; and in several hours it has accomplished, not only the regularity of the general work on all parts of the crystal, but also the re-establishment or regularity in the mutilated part. . ."

"Thus the physical force that arranges the particles of a crystal according to the laws of a wise geometry has results analogous to those which arrange the organized substance in the form of an animal or a plant. This characteristic is not therefore as absolute as Tiedemann thought it to be; nevertheless it has at least a degree of intensity and of energy peculiar to living beings. Moreover, as we have said, there is in the crystal none of the development that characterizes the animal or the plant" (translation by Hoff et al., pp. 25—26).

Bernard returns to the same topic in his Eighth Lecture:

"Morphology is by no means peculiar to living beings, they are not alone in exhibiting specific and constant forms. Mineral substances are capable of crystallizing, these crystals themselves can associate to form diverse and very constant shapes; *clusters, asteroids, macles, prisms,* etc.; at other times the substances take forms that are not truly crystalline, glucose in hillocks, leucine in balls, lecithin in globes, etc.

"There is therefore justification, up to a certain point, for relating the two kingdoms of mineral and living beings, in this sense that we see in the one and in the other this morphological influence that gives a definite form to the part. . .

"These comparisons between mineral forms and living forms certainly constitute only very distant analogies and it would be imprudent to exaggerate them. It suffices to mention them" (translation by Hoff et al., pp. 210—212).

5. Possibility of a crystal as an ancestor of the living organisms

In his book *The Life Puzzle. On Crystals and Organisms and on the Possibility of a Crystal as an Ancestor*, Cairns-Smith [11] considers the connections between "living" and "non-living" forms of the organization of matter, and he considers that a crystal is conceivable as an ancestor of the living organisms, a point to which we shall later return.

REFERENCES

1 T.S. Hall, Ideas of Life and Matter, 2 vol., Chicago and London, 1969.
2 H. Metzger, La Genèse de la Science des Cristaux, Nouveau tirage, Paris, 1969.
3 D.P. Mellor, The Evolution of the Atomic Theory, Amsterdam, 1971.
4 M. de la Métherie, Vues Physiologiques sur l'Organisation Animale et Végétale, Paris, 1780.
5 G. Bachelard, La Formation de l'Esprit Scientifique, Paris, 1947.
6 J.C. Reil, Arch. Physiol., 1759 (reprinted in Sudhoff's "Klassiker der Medizin", Leipzig, 1910).
7 Th. Schwann, Mikroskopische Untersuchungen über die Übereinstimmung in der Struktur und dem Wachstum der Thieren und Pflanzen, Berlin, 1839 (English translation by H. Smith, Microscopical Researches into the Accordance in the Structure and Growth of Animals and Plants, London, 1847).
8 R.C. Maulitz, J. Hist. Med., 26 (1971) 422.
9 Th. Schwann, Leçons d'Anatomie Générale. Texte établi, présenté et annoté par Marcel Florkin, Liège (in preparation).
10 C. Bernard, Lectures on the Phenomena of Life Common to Animals and Plants, vol. I (translation by H.E. Hoff, R. Guillemin and L. Guillemin), Springfield, Ill., 1974.
11 A.G. Cairns-Smith, The Life Puzzle. On Crystals and Organisms and on the Possibility of a Crystal as an Ancestor, Edinburgh, 1971.

Chapter 45

Biosynthesis Considered by Plant Chemists

I. STUDIES IN PHOTOSYNTHESIS

1. Guessing on the basis of organic chemistry

(a) Hexose sugars as first products

The contention of botanists that the starch grains were the first visible products of assimilation (Sachs) had to be taken from a critical viewpoint, as, in a series of papers, Meyer [1] showed that a number of plant species did not contain starch grains (members of *Gentianaceae*, most *Compositae*, *Umbelliferae*, a number of monocotyledons). Chemical analysis showed that, in these plants, sugar is formed; and the analysis of leaves of starch-forming plants also favoured the notion that production of carbohydrate (glucose or other hexose sugars) always precedes the formation of starch. The phenomenon became accepted as an absorption of carbon dioxide in presence of light with a production of carbohydrate and an evolution of oxygen.

Boussingault [2], as well as other authors, starting from the experiments of de Saussure, performed exact determinations of the volume of oxygen evolved and of carbon dioxide absorbed, and he suggested that they are about equal. Assuming the first product to be glucose, or another hexose, the whole process was summed up in the equation

$$6\ CO_2 + 6\ H_2O = C_6H_{12}O_6 + 6\ O_2$$

(b) Intermediate products

If we consider that, in the chloroplast, carbon dioxide is the source of the formation of a hexose molecule, the pathway of this process can in principle be studied experimentally.

Plate 148. Adolf von Baeyer.

Historical studies of this work were accomplished by Schroeder [3], by Stiles [4] and by Rabinowitch [5], and the reader is referred to their publications. The different theories will be described here succinctly, mainly in accordance with the presentation of Stiles.

Liebig, starting from the knowledge that ripening fruits are acid to start with and become sweet, considered that the common plant acids (oxalic, malic, tartaric, citric) are intermediates between CO_2 and glucose. No evidence was obtained to support the idea (see Stiles [4]).

(c) The formaldehyde theory

This theory rests on a suggestion made by the chemist Baeyer [6], on the basis of an observation of Butlerow [7], that trioxymethylene (condensation product of formaldehyde) on heating in alkaline medium gives a syrupy substance with some properties of sugar. The theory has been reigning so long that it is worth quoting its first formulation by Baeyer:

"The general assumption in regard to the formation in the plant of sugar and related bodies is that in the green parts carbon dioxide under the action of light is reduced and by subsequent reactions, transformed to sugar. Intermediate steps have been sought in organic acids: formic acid, oxalic acid, tartaric acid, which can be regarded as reduction products of carbon dioxide. According to this opinion, at those times when the green parts of the plant are most strongly subjected to the action of the sun's rays, a strong accumulation of acids should take place, and these should then gradually give place to sugar. As far as I know, this has never been observed, and when it is remembered that in the plants, sugars and their anhydrides are found under all circumstances, whereas the presence of acids varies according to the kind of plant, the particular part and its age, then the opinion already often put forward, that the sugar is formed directly from the carbon dioxide, increases in probability.

"The discovery of Butlerow provides the key, and one may indeed wonder that so far it has been so little utilized by plant physiologists. The similarity which exists between the blood pigment and the chlorophyll has often been referred to; it is also probable that chlorophyll as well as haemoglobin, binds carbon dioxide. Now, when sunlight strikes chlorophyll which is surrounded by CO_2, the carbon dioxide appears to undergo the same dissociation, oxygen escapes, and carbon monoxide remains bound to the chlorophyll. The simplest reduction of carbon dioxide is that to the aldehyde of formic acid; it only requires to take up hydrogen,

$$CO + H_2 = COH_2$$

References p. 172

This aldehyde is then transformed, under the influence of the cell contents as well as by alkalies, into sugar. As a matter of fact, it would be difficult, according to the other opinion, by a successive synthesis, to reach the goal so easily! Glycerol could be formed by the condensation of three molecules, and the subsequent reduction of the glyceric aldehyde so formed" (translation by Jörgensen and Stiles [8]).

(d) Forms of the formaldehyde theory

After being formulated as a suggestion by Baeyer in 1870, the formaldehyde theory took a number of different forms. A direct reduction of carbonic acid to formaldehyde was proposed by Reinke [9]; whereas Maquenne [10] suggested methane as an intermediate between carbonic acid and formaldehyde. According to Bach [11], hydrogen peroxide was also formed; and it was, according to Usher and Priestly [12] removed by catalase.

Hydrogen was considered by several authors as the effector of the reduction of carbon dioxide. Several schemes were proposed to account for the origin of the hydrogen involved: from organic compounds (Pollacci [13,14]); from the splitting of water under the action of light (Kimpflin [15]); or from the enzymatic destruction of carbohydrates (Stoklasa and Zdobnicky [16]).

2. Evidence in favour of the theory, based on the formation of formaldehyde in different model systems

(a) Systems containing carbon dioxide and water

Formaldehyde is, for example, produced by a reduction of carbon dioxide by magnesium (Fenton [17]) or by silent electric discharge (Löb [18]). It has also been claimed to result from the action of ultraviolet rays (Berthelot and Gaudechon [19], Usher and Priestley [20], Stoklasa and Zdobnicky [16]).

(b) Systems containing carbon dioxide, water and chlorophyll

The formation of formaldehyde in such systems was claimed by Pollacci [13], Usher and Priestley [12], Schryver [21], Chodat and Schweizer [22]. Warner [23] and Ewart [24] found no trace of formaldehyde production under such conditions. On the other hand, Warner [23] and Ewart [24] observed a production of formaldehyde in mixtures of chlorophyll, water and oxygen. These results suggested that the formation of formaldehyde considered to take place in the experiment reported above may have been due to an action of oxygen on chlorophyll. This was confirmed by subsequent experiments showing that in systems containing only carbon dioxide, water and pure chlorophyll, no formaldehyde is produced (Jörgensen and Kidd [25]).

3. Biochemical approaches based on the formaldehyde theory

(a) Search for formaldehyde in green leaves

As stated by Rabinowitch [5]:

"Because of the popularity of Baeyer's "formaldehyde theory" (1870) no other compounds has been so eagerly searched for in plants as formaldehyde and with such uncertain results. Categorical statements that formaldehyde does occur in leaves have been answered by no less categorical denials."

A number of statements about the presence of formaldehyde in illuminated leaves are based on direct determinations by more or less specific methods or reagents.

In the products of leaf distillation, Reinke [26] recognized the presence of an aldehyde he thought to be formaldehyde, but Curtius and Reinke [27] showed that it lacked the specific properties of formaldehyde. Distillates of green leaves were claimed by Pollacci [28] to present a positive formaldehyde test but the results were opposed by Plancher and Ravenna [29]. A new reagent, diphenylamine and sulphuric acid, was introduced by Grafe [30] who claimed to have observed positive results, but Curtius and Franzen [31] showed the reagent to lack specificity for formaldehyde.

(b) Feeding plants with formaldehyde

Claims have been formulated by different authors about the formation of carbohydrates in plants after they had been fed with formaldehyde or derivatives of it. Loew [32] and his collaborator Bokorny [33] claimed that starch is formed in *Spyrogyra* from the sodium bisulphite compound of formaldehyde. On the other hand, several authors have claimed that free formaldehyde can be assimilated by the following plants: the alga *Trichoderma viridis* (Boitreux [34], Moore and Webster [35]), and higher green plants (Baker [36]; Grafe and Vieser [37]).

(c) Alleged photosynthesis in vitro

Experiments in vitro, the results of which were published between 1921 and 1929 by Baly [38—40], of Liverpool, and which nobody has been able to reproduce, led him to claim that, when solutions of CO_2 are exposed to ultraviolet radiation, appreciable amounts of formaldehyde are produced; and further that, if a coloured catalyst, such as cobalt and nickel carbonate, is added, the process takes place in ordinary light with a production of reducing sugar. The subject gave rise to a number of publications. (Literature in Rabinowitch, vol. I.) These views, presented in the form of a book by Baly [41], as late as 1940, now have no credence among researchers in the field of photosynthesis (see Stiles [4]). Curiously, the concept still appears in texts concerned with the particular biosynthetic process considered in the context of the origin of life *. On Baly's views, J.B.S. Haldane [42] based his theory of chemical evolution before the origin of life expressed in the following sentence, still sometimes quoted.

> "Now, when ultra-violet light acts on a mixture of water, carbon dioxide and ammonia, a vast variety of organic substances are made, including sugars and apparently some of the materials from which proteins are built up. This fact has been demonstrated in the laboratory of Baly of Liverpool and his colleagues. In this present world, such substances, if left about, decay — that is to say, they are destroyed by microorganisms. But before the origin of life they must have accumulated till the primitive oceans reached the consistency of hot dilute soup" (Haldane [42]).

It is interesting to note that a notion, which has been discarded as unsubstantiated in one chapter of a science, persisted among the

---

* It will not be dealt with here, as the present author has contributed a historical essay on the subject in vol. 29B (Chapter V) of this Treatise.

shibboleths of another chapter where it became the basis of another, even more persistent shibboleth, the "hot soup".

(d) Experiments with dimedon

The reagent dimedon (dimethylhydroresorcinol) was successfully used by Neuberg in studies on the pathway of glycolysis, as a test for aldehydes. Klein and Werner [43] claimed in 1926, to have demonstrated the production of formaldehyde, isolated in the form of formaldomedon. These experiments were most extensively conducted with aquatics, especially *Elodea canadiensis*, and the authors ascertained that formaldomedon (which they identified in the external medium) was only produced in the light and in the presence of carbon dioxide. For a few years, these results appeared to some as convincing evidence for the validity of the formaldehyde theory until Barton-Wright and Pratt [44] showed that there was no experimental foundation for Klein and Werner's statement.

4. Theories inspired by biocolloidology

The present author has proposed a definition of what he has designated, in Chapter 14, by the term "biocolloidology". This theory claimed that "protoplasm" consisted of different and variable aggregation states (micelles) of compounds. The biological phenomena were, according to the theory, the effects of surface phenomena resulting from changes in agglutination, dispersion, hydration or dehydration of the micelles, resulting from the action of ions. Energy in organisms, was traced back to energies of surface films and interfaces. The author wished to reiterate that, by recognizing the retarding effect of biocolloidology as understood above, he has no intention of deprecating colloid chemistry, a respectable chapter of chemistry which has led to a number of useful applications in the fields of analytical chemistry, pharmacy, industrial chemistry, etc.

Biocolloidology, which started as far back as 1861 with the formulation by Graham [45] that

"The colloid possess *ENERGIA*. It may be looked upon as the probably primary source of the force appearing in the phenomena of vitality"

References p. 172

became widely spread, in the beginning of this century and knew its apex in the twenties. It did not fail to influence for some time the theories of photosynthesis.

As chlorophyll is contained in chloroplasts, it was agreed at the time that photosynthesis took place in these organelles. In agreement with the current theories, Willstätter [46] considered that chlorophyll was dispersed "as a colloidal hydrosol" and "adsorbed" at a surface where the photochemical action took place. Stern [47] claimed that the chloroplast consisted of an emulsion colloid, a lipoid phase of which is distributed through an aqueous-protein phase, the chlorophyll being in solution in the lipoid phase. The reactions of photosynthesis, according to the biocolloidal theory formulated by Stern, take place at different sites, one part in one phase and another in another phase, the surface between these phases playing the essential role. A blocking of the surface, resulting from the accumulation, at this surface, of active substances such as phenylurethane, would retard the process of photosynthesis.

Warburg [48] had already, before Stern, proposed that the action of narcotics on photosynthesis was an action on a limiting surface. Though not indulging in the sweeping statements of biocolloidology, Warburg, as was stated in Chapter 27, started with a model inspired by surface phenomena and, in studies performed in whole cells and models, he first considered the participation of a ferric iron colloidal complex. *Post hoc* is not, nevertheless, *propter hoc*; and it was Warburg himself who characterized his *Atmungsferment* as a well-defined enzyme (see Chapter 27, p. 209) and recognized it as a heme compound. We have recalled how he became converted to the dehydrogenase concept in 1929 (Chapter 36, p. 370). It was after a rupture with biocolloidology, and not as a sequence of this, that Warburg accomplished his main contributions to biochemistry.

II. AMINO ACIDS RECOGNIZED AS PRECURSORS OF PROTEINS AND OF AMIDES IN AN APPROPRIATE BIOLOGICAL SYSTEM (PLANT SEEDLINGS)

This section concerns the early studies on protein biosynthesis, accomplished by plant chemists on plant seedlings. For a period, this material was as widely used as *E. coli* and the Wistar rat are nowadays. The abundant literature on the subject has been clarified by Chibnall * in Chapter I—IV of his book *Protein Metabolism in the Plant* [49], based on his Silliman Lectures. This analysis, on which the following text is based, remains a classic of the History of Biochemistry.

1. Asparagine in the seedlings of Papilionacea

Asparagine was observed, in 1806, by Vauquelin and Robiquet [51] to crystallize out of a large amount of juice of *Asparagus* shoots. They recognized asparagine as a neutral nitrogen-containing compound. From it, Plisson [52] isolated what was later called aspartic acid. That asparagine was according to our language, an amide of aspartic acid was recognized by Pelouze [53] in 1833, though the constitution of the latter was only established

* It is not uncommon for a scientist to become a historian of his science, but the case of Chibnall, who has become a general historian is more remarkable. Chibnall's family originated from the village of Sherington in North Buckinghamshire. He had his first taste of historical research when, at the time of his return from World War I, he found among family papers the marriage settlement of his grandfather's grandfather, drawn up in 1760 and dealing with land in Sherington. At this time, to ferret out information, Chibnall visited the Public Record Office where he found such a wealth of information that he soon knew quite a lot about Sherington's history. While pursuing his career as a scientist he resolved that he would one day return to these fascinating historical studies. After a long and brilliant career on plant protein biochemistry (see ref. 50) he succeeded F.G. Hopkins in Cambridge in 1943. He resigned in 1949 from this charge, for the reasons he has explained [50], and returned to his work on proteins and waxes which he pursued until 1957.

"By now I felt it was time to take stock, for the three score years and five were just round the corner." (Chibnall was born in 1894.)

Chibnall, since then, has been active as a historian.

"Sherington, where my forebears had resided, emerged once more as a focus of study, and the interpretation of historical evidence for the feudal and economic development of the countryside became as exciting a challenge as the analysis of proteins and waxes had been during my more active years as a biochemist" [50].

In 1965, Chibnall published an important book on the History of Sherington, and a second major historical work of his has just gone to the press. (See also Notes added in proof.)

Plate 149. Wilhelm Pfeffer.

with certainty by Kolbe [54] in 1862. The β-amide nature of asparagine was established in 1887 by Piutti [55], who accomplished its synthesis.

The presence of large amounts of asparagine in plant seedlings was observed by a professor of the University of Pisa, Piria. As told by Chibnall [49] (p. 2), a pharmacist of Pisa, Menici, had raised vetch seedlings in the dark and, from their extract, he had prepared a crystalline material which he submitted to Piria for analysis. Piria [56,57] not only recognized asparagine but observed that he could isolate large amounts of it whether the seedlings were kept in the light or in the dark. As he could isolate free asparagine neither from ungerminated seeds nor from adult plants bearing fruit, he astutely concluded that asparagine was produced from a nitrogenous reserve (what he called a "casein") and that it disappeared when the plant reached the adult stage. In 1848, Dessaignes and Chautard [58] confirmed the accumulation of asparagine in vetch seedlings grown both in light and dark, as well as in a series of other leguminous species, but they did not find it in buckweat, pumpkin or oats even when grown in the dark.

It may be remembered, in the history of asparagine, that in the early 1850's the classical studies of Pasteur [59] on the properties of malic acid and of aspartic acid took place. Piria had observed that aspartic acid is converted, under the action of nitric acid, into malic acid, and Dessaignes had prepared aspartic acid by heating the ammonium salts of malic or fumaric acid. Pasteur showed that this aspartic acid prepared by Dessaignes had no optical activity, whereas that obtained from biological sources was optically active.

2. Asparagine considered as a direct primary product of the oxidation of protein reserves

In 1872, Pfeffer [60], in his elaborate research on the genesis of proteins in grain ripening and germination, expressed his conviction, based on a large number of observations, that the asparagine so abundantly found in vetch seedlings was a direct primary product of the oxidation of the protein reserve of the seeds, an idea that had been suggested by Boussingault [61], without experimental basis, in 1868. Pfeffer considered that the asparagine

References p. 172

formed from proteins in cotyledons diffused out into the axial organs where, if carbon and hydrogen were available from such sources as carbohydrate, the asparagine derived from the seed protein, legumin, would be reconverted by a direct combination with carbohydrate, into another protein, albumin.

It must be noted that Pfeffer, to establish this theory of protein biosynthesis, performed no chemical analysis and relied on old microchemical methods familiar to botanists. It must also be noted that, at the time of Pfeffer, only scanty knowledge was available about the nitrogenous constituents of plants. Asparagine was conspicuously abundant in some plant seedlings while the other non-protein nitrogenous constituents of plants known at the time, nitric acid and ammonia, were considered as not likely to derive from seed proteins by oxidation.

3. Recognition of aspartic acid as a constituent of plant proteins

The preparation of aspartic acid as a product of the acid hydrolysis of conglutin, a lupin protein, by Ritthausen [62] in 1869 was considered by Pfeffer as supporting his ideas. In fact, it was in itself a landmark in protein chemistry. To make precise the situation with respect to the origin of conceptual relations between the nature of "albuminous compounds" on the one hand, and amino acids on the other hand, it is interesting to state that the concept of the compounds which were called amido acids and later amino acids originated without reference to "albuminous compounds". Leucine, α-aminoisocaproic acid,

$$(CH_3)_2CHCH_2\underset{\underset{NH_2}{|}}{C}HCOOH)$$

had been isolated in the free form from cheese by Proust [63] in 1819. On the other hand, in 1820 glycine (amino acetic acid)

$$\underset{\underset{NH_2}{|}}{C}H_2-COOH$$

was isolated from a sulphuric acid hydrolysate of gelatine by Braconnot [64]. Braconnot was unaware of the presence of nitrogen in the compound which he called glycocoll on account of its sweet taste (sucre de gélatine, sugar of gelatine) and which was later called glycine *.

In 1848, Laurent and Gerhardt recognized a chemical relationship of leucine and glycine, and these chemists proposed to recognize these two compounds as members of the same homologous series, $C_n H_{2n} + 1 NO_2$.

In 1860, when Ritthausen began his studies on plant proteins, only four amino acids had been identified by animal chemists among protein derivatives; and among those, only leucine and tyrosine were considered as widely spread, while glycine and serine were considered as limited to special sources.

Leucine, which had first been isolated from cheese as stated above, was isolated from several proteins (muscle, wool) by Braconnot [64] in 1820. L-Tyrosine had been obtained by Liebig [65], in 1846, from an alkaline hydrolysate of casein. Bopp [66], in 1849 had isolated it from several animal proteins (albumin, casein, fibrin). The synthesis by Erlenmeyer and Lipp [67], in 1883, established the structure.

Another amino acid that had been isolated from a particular source, silk, in 1865 was L-serine (α-amino-β-hydroxypropionic acid) obtained by Cramer [68], who recognized its amino acid nature. (Fischer and Leuchs [69] were to establish the structure by synthesis in 1902.)

In 1866, Ritthausen [70], from the plant protein now designated as wheat gliadin, obtained a new amino acid, L-glutamic acid (α-aminoglutaric acid)

$$\begin{array}{l} COOH \\ | \\ CH_2 \\ | \\ CH_2 \\ | \\ CHNH_2 \\ | \\ COOH \end{array}$$

In 1868, Ritthausen [71] hydrolyzed the chief reserve protein of lupin seeds, conglutin, and, besides leucine, tyrosine and glutamic acid, obtained a new amino acid. This amino acid he recognized in 1869 [62] as being aspartic acid, already known as a product of asparagine hydrolysis.

* Glycine has been adopted only recently in French literature, as in French the word already commonly described a flower (*Wistaria*).

References p. 172

Plate 150. Heinrich Ritthausen.

4. First suggestions of the presence of amino acid as such in the structure of albuminous compounds

At a time when few scientists had any idea of the significance of amino acids as protein constituents, these discoveries of Ritthausen must be considered as highlights in the history of protein chemistry. Another landmark was the publication, in 1872, of Ritthausen's book *Die Eiweisskörper* [72]. Considering the content of several plant proteins in glutamic acid and in aspartic acid, Ritthausen formulated the notion that this content was characteristic of the protein concerned.

Another point underlined by Ritthausen was that the aspartic acid isolated from proteins (as well as from asparagine, contrary to the synthetic aspartic acid which, as shown by Pasteur, had no optical activity) was optically active, from which he concluded that the protein itself must contain aspartic acid as such. So far, the possibility of a formation of amino acids during the process of isolation by acid or alkaline hydrolysis had not been discarded.

As Chibnall [49] states (p. 14):

"Ritthausen was voiding here for the first time two important postulates:

(*1*) that the amino acids produced on hydrolysis were derived from structures already formed in the protein molecule, a question which was to be hotly disputed by chemists during the ensuing 20 years, and

(*2*) that asparagine was probably present in the protein molecule".

5. Arguments in favour of the presence of amides in proteins

As noted by Chibnall [49] (p. 14), indirect support was furnished during the same year, 1872, by O. Nasse [73—75] in favour of the presence not only of aspartic acid, but also of asparagine within protein molecules. Otto Nasse has furnished original contributions to many fields: he demonstrated, for instance the normal presence of glycogen in muscle. He was the son of the physiologist Henrich Nasse and he was professor of pharmacology and physiological chemistry in Rostock. He worked out an exact method for the determination of the ammonia liberated from proteins during their hydrolysis, and he observed that the amount of ammonia liber-

References p. 172

Plate 151. Albert Charles Chibnall.

ated in the course of alkaline hydrolysis was higher than in acid hydrolysis. The rate of liberation was rapid at first and slower later. From his experiments, Nasse inferred that he could distinguish "firmly bound ammonia" (later shown to originate from arginine) and "loosely bound ammonia". The latter, among other possibilities, might be present in the protein molecule in the form of acid amide groups. He instanced the like behaviour of asparagine.

This suggestion of Nasse was followed up by Hlasiwetz and Habermann [76]. They hydrolysed casein with hydrochloric acid in the presence of tin and obtained leucine, tyrosine, aspartic acid, glutamic acid and ammonia. They suggested that the ammonia is combined partly with aspartic acid as was known for asparagine and partly with glutamic acid in the form of a still unknown amide which they proposed to call glutamine. This suggestion, formulated in 1873, was confirmed a decade later by the isolation, from beet root, of glutamine by Schulze and Bosshard [77]. The prediction of Hlasiwetz and Habermann was accomplished by Chibnall and his collaborators [78] who isolated glutamine from an enzymatic hydrolysate of gliadin in 1932. A year later, the first synthesis of glutamine was accomplished by Bergmann, Zervas and Salzmann. On the other hand, asparagine was isolated from the protein edestin by Damodaran [80] in 1932.

Returning to the important paper of Hlasiwetz and Habermann [76] (1873), we note that these authors believed that casein, on acid hydrolysis yields almost exclusively leucine, tryosine, aspartic acid, glutamic acid, and as they suggested, asparagine and glutamine. But, as stated by Chibnall [49] (p. 15), it appears that Schulze, more than most of his contemporaries, believed that the amino acids detected by Hlasiwetz and Habermann represented only a part of the protein molecule. The notion that holoproteins are composed exclusively of amino acids came only later, and other compounds were at the time regarded as possible constituents of proteins, as for instance xanthine and hypoxanthine, isolated by Salomon [81], from lupin seedlings (see Chapter 49, section 5).

Plate 152. Eugen Franz von Gorup-Besanez.

6. Enzymatic liberation of amino acids in seedlings and their use for protein biosynthesis as suggested by Gorup-Besanez

In 1874, Gorup-Besanez, professor of chemistry at Erlangen, published a first paper on the studies of vetch seeds and seedlings which occupied him during the last years of his life. Gorup-Besanez, who died in 1878, was a well-known analyst in the field of animal chemistry, and the renowned author of widely read treatises. He had [82], in 1856, discovered α-aminoisovaleric acid, later called valine,

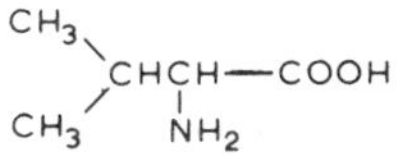

$$(CH_3)_2CH{-}CH(NH_2){-}COOH$$

in extracts of pancreas. (It is only after the death of Gorup-Besanez that Schützenberger [83] recognized valine as a constituent of albumin, in 1879.)

The acute mind of Gorup-Besanez recognized the possibilities of situating a biochemical process, protein biosynthesis, in its appropriate biochemical context, by turning his analytical skill to the seedlings to which his attention had been called by Pfeffer's theory of protein "regeneration".

In 1874, Gorup-Besanez [84] announced that he had found that etiolated vetch seedlings, 2—3 weeks old, contained not only asparagine, but also leucine (the product he described was later shown to be a mixture of leucine and phenylalanine). During the same year, Gorup-Besanez [85] stated that, as he could find no leucine in ungerminated seeds, its presence in seedlings proves that it had been liberated during germination. He [86] isolated from the seeds a "ferment" transforming albumin into peptone. In 1877, Gorup-Besanez [87], having detected glutamic acid and tyrosine in vetch seedlings concluded that during the germination, a splitting of the albuminous compounds of the seed reserve takes place, analogous to the splitting which is undergone by such substances in animal digestion, and corresponding to the splitting of animal albuminous compounds observed by Hlasiwetz and Habermann in alkaline hydrolysis. In fact he had isolated the four protein amino acids obtained by these authors.

At a time when enzymes were still considered as limited to the

References p. 172

Plate 153. Ernst Schulze.

preparatory digestive process taking place in the digestive tract of higher animals, the discovery by Gorup-Besanez of an enzymatic production of amino acids used in protein biosynthesis from an albuminous compound in a plant seedling was widely recognized as a highlight.

7. Schultze's contributions

That the seed reserve proteins are enzymatically decomposed into amino acids and amides during germination, as suggested by Gorup-Besanez, was documented by the extensive studies of E. Schulze. A student of Wöhler in Göttingen, Schulze was appointed as professor of agricultural chemistry at the Polytechnicum of Zürich in 1872. Schulze recognized the importance of the suggestion of Gorup-Besanez. He noted in 1878 [88] that during the germination of lupin seeds, an increase of amino-nitrogen * takes place, and he actually isolated asparagine, leucine and tyrosine. In pumpkin seedlings, Schulze and Barbieri [89] found leucine, tyrosine, asparagine and glutamine (indirectly detected as glutamic acid). The same authors [90], in 1879, isolated phenylalanine from lupin seedlings. They obtained this amino acid from hydrolysates of conglutin, a protein of the lupin seed, in 1881 [91]. Owing to the chemical synthesis of phenylalanine accomplished by Erlenmeyer and Lipp [92] in 1882, Schulze and Barbieri established the identity of the natural amino acid with the synthetic one.

In lupin seedlings, valine was found in 1883 by Schulze and Barbieri [93].

Glutamine was, as stated above, obtained from beet juice in

* As noted by Chibnall [49] (p. 19, footnote), no distinction was made at the time between amino acids and amino acid-amides, their nitrogen being indiscriminately referred to as "amido-N". Nevertheless in Schulze's earlier papers, although he uses the term "amides" for amino acids, he uses "amido-N" with the meaning of our "amino-N", though it was not general practice, some authors using "amides" to designate non-protein nitrogenous products. Chibnall considers that the modern expression "amino-N" came into general use with its present significance after Van Slyke described his estimation method.

References p. 172

1882 by Schulze and Bosshard [77]. As stated above, the views of Pfeffer could be epitomized by stating that in the lupin seedlings, the reserve proteins of the seed were broken down into sugar and asparagine and that these two substances diffusing into the axial organs were the precursors of the proteins biosynthesized there. This Schulze was not able to accept. In a paper of Schulze and Umlauft [94] published in 1876 he suggested that the breakdown of the reserve protein gave not only asparagine but also amino acids which were, more than asparagine, likely precursors of the proteins of the axial organs.

It must be noted that if much progress had been done concerning the nature of the decomposition of proteins, the constitution of the protein molecule remained unknown (the peptide bond concept was formulated later, independently, by Hofmeister and by E. Fischer; see Chapter 15) Schulze firmly believed in the concept accepted by Gorup-Besanez, according to which the amino acids isolated from proteins were preexisting in their molecules and he consequently was in favour of a biosynthesis of proteins by an association of amino acids even if at the time the whole collection of these amino acids had not yet been completely catalogued (see section 10 of the present Chapter). Baumann [95] had, in 1882, shown the preexistence in the protein molecules, of certain atomic groupings observed in their decomposition products. The Millon test for tyrosine was given by proteins. Oxidation gave benzoic acid and benzaldehyde in sufficient amounts to indicate the presence of another aromatic group, and it was known that besides tyrosine, phenylalanine contained such a group.

But there were opponents. For instance, Löw [96] claimed that, as proteins do not reduce osmic acid while leucine does it slowly, he could not accept the presence of leucine as such in proteins. But, Baumann retorted, glucose and fructose both reduce Fehling's solution, while saccharose does not. Later, Löw [97] admitted that both aromatic and phenolic nuclei might be present in the protein molecules but he maintained his opposition to the presence of leucine, as he could not obtain valeric acid by oxidation.

In 1878, Schulze [88] argued that if, as Gorup-Besanez suggested, the seed reserve proteins are enzymatically hydrolysed into what we now call amino acids and amides, the relative proportions of these should correspond to the relative proportions in acid or alkaline hydrolysates. But this is not confirmed. For instance, conglutin of lupin seeds yields much leucine while this amino acid is found in very small proportions in seedlings. Conglutin yields, on hydrolysis, more glutamate than aspartate: more glutamine than asparagine should therefore be expected in seedling. The contrary is observed. If we consider that the proteins synthesized have a composition which differs from that of reserve proteins the observed facts become clear. Asparagine, in lupin seedlings (or glutamine in pumpkin seedlings) would belong to the protein constituents slowly utilized. But this would not explain the considerable accumulation of asparagine in lupin seedlings: this must come from a synthesis at the expense of other decomposition products.

In 1880 Pfeffer published his *Handbuch der Pflanzenphysiologie* [98] where he opposed the views of Gorup-Besanez and of Schulze regarding the hydrolysis of reserve proteins, yielding their amino acids and amides. It must be granted that, at the time, there were facts in support of Pfeffer's view who claimed that the relative proportions of decomposition products varied from one method to another. For instance valine had not been recognized as a product of protein acid hydrolysis, while, as stated above, Schützenberger [83] discovered it in alkaline hydrolysates. Another example was the difference of proportions of aspartic acid obtained from conglutin after acid hydrolysis by Ritthausen [72] and by Hlasiwetz and Habermann [99] after treatment of the protein with bromine. These contradictions are explainable for us, but at the time it was natural that they may have influenced biologists. In 1885, Schulze [100] replied that the task of the plant chemist was to identify the chemical processes taking place in organisms with chemical processes that can be accomplished outside of it and he referred to the chemical evidence in favour of the presence of amino acids in protein molecules (see above).

The accumulation of asparagine in plant seedlings he refused to consider as any aspect of "protoplasmic theory" and he considered that improvements were to be pursuant to studies of protein constitution and the exploration of non protein constituents of plants. He contributed to this by his discoveries related above, of phenylalanine, of valine and of arginine in seedlings.

Schulze [101] epitomized in 1906 his views on protein synthesis and on asparagine synthesis, resulting from his long activity in identifying amino acids and other nitrogenous products in plant

References p. 172

Plate 154. Dimitri Nicolaevich Prianishnikov.

organs. He reports on the presence in seedlings of ten of the amino acids known as protein hydrolysis products: valine, leucine, isoleucine, phenylalanine, tyrosine, tryptophan, proline, arginine, histidine and lysine. These were found only after germination and were recognized as derived from seed proteins. He isolated many other nitrogenous compounds. He had also found asparagine and glutamine but as stated above, it was only in 1932 that these amides were demonstrated to be protein constituents.

In keeping with his view that asparagine was elaborated from amino acids he provided the proof that the nitrogen of asparagine originated from the catabolism of amino acids. In 1899, Schulze's student Prianishnikov [102] had clearly proved that part of the asparagine must come from protein decomposition. He thought that the remaining part derived from amino acids. In 1899, when less than 50 per cent of the components of proteins were accounted for as amino acids there was an element of uncertainty as it was not yet evident that proteins would give only amino acids on enzymatic, or even acid, hydrolysis.

8. Glutamine as secondary product of seedling metabolism

Protein metabolism, in many seedlings, for instance of *Cruciferae* leads to an accumulation of glutamine instead of asparagine as it is the case in Leguminous plants (see Stieger [103])

That, at least partly, this amide is formed by synthesis and is a secondary product of protein metabolism has since been documented (see Chibnall [49], p. 60).

9. Amide biosynthesis

Schulze's work had rendered untenable the theory of Pfeffer, according to which the whole of the amides arose directly from proteins. The view prevailed that amides accumulated in plant seedlings were partly formed in a secondary process, by a biosynthesis involving ammonia derived from amino acids.

When Prianishnikov, in 1895, returned to the Petrowsko-Rasumowskoje Academy of Moscow, he pursued the trend of research

References p. 172

of his master Schulze. He conceived the idea (first opposed by Schulze) that the conversion of amino acids into ammonia was not essentially the first stage of the process of protein "regeneration", but a general metabolic process. He stated this concept in 1899 in a paper [102] in which he foresaw the general metabolic signification of deamination which was shortly afterwards to be developed by animal chemists, as was retraced in Chapter 35.

Prianishnikov recalled the view of Boussingault who thought that the production of asparagine in plants was analogous to the production of urea in animals, both being a result of the "burning up" of proteins. Prianishnikov concluded that the protein "regeneration" in plant seedlings was accomplished at the expense of amino acids and that asparagine and glutamine, contributing to the process, were for their greatest part by-products arising from a secondary conversion of amino acids. Schulze was convinced of the production of ammonia by the breakdown of amino acids when Butkewitsch [104], in 1909, showed that, in plant seedlings anaesthetized with toluene vapour, synthetic processes ceased, but protein decomposition went on, liberating ammonia, which accumulated instead of asparagine. Schulze remained unconvinced that the process was an oxidation, which was demonstrated by Krebs in 1933 (see Chapter 35, p. 326). In 1924, Prianishnikov, whose work reinforced the concept of amide biosynthesis in plants as situated in a metabolic prolongation of the nitrogen catabolism of amino acids, formulated a physiological parallel between asparagine biosynthesis in plants and urea formation in animals, both considered by him as biosynthetic processes involved in ammonia detoxification.

10. Amino acids as constituents of proteins

The history of their discovery was been retraced by Vickery and Schmidt [105], by Meister [106] and by Fruton [107] (pp. 108—111). We may note that the list of the amino acids recognized as protein constituents, as partly retraced in Fig. 4 of Chapter 15, attained its maximum at the end of the period considered in the present volume.

When, at the end of the 19th Century, Emil Fischer became interested in proteins, the amino acids which had been recognized as cleavage products of proteins amounted to 13. Some of these were discovered among products of protein hydrolysis at the following dates: glycine (1820), leucine (1820), tyrosine (1846), serine (1865), glutamic acid (1866), aspartic acid (1869), valine (1879), phenylalanine (1881).

At the time, other amino acids had been recognized as protein constituents. Alanine (α-amino propionic acid) had been prepared by synthesis by Strecker [108] (1850) before its recognition as a natural substance (isolated from silk by Weyl [109] in 1888). Lysine was isolated from wool by Drechsel [110] in 1889.

Histidine was isolated by Kossel [111] from sturine, a protamine of sturgeon sperm, in 1896. Arginine was isolated from hydrolysates of horn by Hedin [112] in 1895. Cystine, which had first been obtained from a urinary calculus (whence its name) by Wollaston [113] in 1810 was isolated from a hydrolysate of horn by Mörner [114] in 1899. That cysteine was a disulfide which could be reduced to the corresponding mercaptan, cysteine, was shown by Baumann [115] in 1884 and later recognized as a protein constituent.

After the turn of the century, more amino acids were added to the list of protein constituents. Proline, which had been synthesized by Willstätter [116] in 1900 was recognized by Fischer [117] in hydrolysates of casein in 1901.

The following year, Fischer [118] isolated hydroxyproline from gelatin. In 1901, Hopkins and Cole [119] obtained the indole-containing amino acid tryptophan from casein. They were led to this discovery by trying to isolate from protein hydrolysates the material producing the violet color first observed by Adamkiewicz [120] (1874), when adding sulfuric acid to a mixture of glacial acetic acid and albumin. Isoleucine was isolated by F. Ehrlich, [121] in 1907 from hydrolysates of fibrin, wheat gluten, egg albumin and beef muscle. He had already, in 1904 [122] isolated it from beet sugar molasses. Mueller [123] isolated methionine from casein in 1922. The last discovery of a protein amino acid dates to 1935, with threonine, isolated from fibrin by Rose et al. [124]. As was stated above, asparagine and glutamine were identified as protein constituents in 1932.

References p. 172

REFERENCES

1 A. Meyer, Bot. Z., 43 (1885) 417, 433, 449, 465, 481, 497.
2 J.B. Boussingault, Ann. Sci. Nat., Bot., 5th Ser., 1 (1864) 31.
3 H. Schroeder, Die Hypothesen über die chemische Vorgänge bei der Kohlensäureassimilation und ihre Grundlagen, Jena, 1917.
4 W. Stiles, Photosynthesis, London, 1925.
5 E.I. Rabinowitch, Photosynthesis and Related Processes, vol. I, New York, 1945.
6 A. Baeyer, Ber. Dtsch. Chem. Ges., 3 (1870) 63.
7 A. Butlerow, Ann. Chem., 120 (1861) 295.
8 I. Jörgensen and W. Stiles, Carbon Assimilation: a Review of Recent Work on the Pigments of the Green Leaf and the Processes Connected with Them, London, 1917.
9 J. Reinke, Lehrbuch der allgemeinen Botanik mit Einschluss der Physiologie, Berlin, 1880.
10 L. Maquenne, Bull. Soc. Chim., Paris, N.S., 37 (1882) 298.
11 A. Bach, Compt. Rend. 116 (1893) 1145.
12 F.L. Usher and J.H. Priestley, Proc. Roy. Soc., Ser. B, 77 (1906) 369.
13 G. Pollacci, Atti Ist. Bot. Univ. Pavia, N.S., 8 (1902) 1.
14 G. Pollacci, Arch. Ital. Biol., 36 (1902) 446.
15 G. Kimpflin, Essai sur l'Assimilation Photochlorophyllienne du Carbone, Thesis, Lyon, 1908.
16 J. Stoklasa and W. Zdobnicky, Biochem. Z., 30 (1911) 433.
17 H.J.H. Fenton, J. Chem. Soc., Trans., 91 (1907) 687.
18 W. Löb, Landw. Jahrb., 35 (1906) 541.
19 D. Berthelot and H. Gaudechon, Compt. Rend. 150, (1910) 1690.
20 F.L. Usher and J.H. Priestley, Proc. Roy. Soc., Ser. B., 84 (1911) 101.
21 S.B. Schryver, Proc. Roy. Soc., Ser. B., 82 (1910) 226.
22 R. Chodat and K. Schweizer, Arch. Sci. Phys. Nat., 4e Pér., 39 (1915) 334.
23 C.H. Warner, Proc. Roy. Soc., Ser. B., 87 (1914) 378.
24 A.J. Ewart, Proc. Roy. Soc., Ser. B., 89 (1915) 1.
25 I. Jörgensen and F. Kidd, Proc. Roy. Soc., Ser. B., 89 (1915) 342.
26 J. Reinke, Ber. Dtsch. Chem. Ges., 3 (1870) 63.
27 Th. Curtius and J. Reinke, Ber. Dtsch. Chem. Ges., 15 (1897) 201.
28 G. Pollacci, Atti Ist. Bot. Univ. Pavia, 6 (1899) 27.
29 G. Plancher and C. Ravenna, Atti Accad. Lincei, 13 II (1904) 459.
30 V. Grafe, Oesterr. Bot. Ztg., 56 (1906) 289.
31 Th. Curtius and H, Frazen. Ber. Dtsch. Bot. Ges., 45 (1912) 1715.
32 O. Loew, Ber. Dtsch. Chem. Ges., 22 (1889) 482.
33 T. Bokorny, Ber. Dtsch. Chem. Ges., 24 (1891) 103.
34 Boitreux, C.R. Soc. Biol., 83 (1920) 737.
35 B. Moore and T.A. Webster, Proc. Roy. Soc., Ser. B., 91 (1920) 196.
36 S.M. Baker, Ann. Bot., 27 (1913) 411.
37 V. Grafe and E. Vieser, Ber. Dtsch. Bot. Ges., 27 (1909) 431, 29 (1911) 19.

38 E.C.C. Baly, J. Chem. Soc., 119 (1921) 1025.
39 E.C.C. Baly, Proc. Toy. Soc., Ser. A, 116 (1927) 212.
40 E.C.C. Baly, Proc. Roy. Soc., Ser. A, 122 (1929) 393.
41 E.C.C. Baly, Photosynthesis, New York, 1940.
42 J.B.S. Haldane, Rationalist Annual, (1929) 148.
43 G. Klein and O. Werner, Biochem. Z., 168 (1926) 361.
44 E.C. Barton-Wright and M.C. Pratt, Biochem. J., 24 (1930) 1210.
45 T. Graham, Phil. Trans. Roy. Soc. (London), (1861) 183.
46 R. Willstätter, Ber. Dtsch. Chem. Ges., 35 (1922) 3601.
47 K. Stern, Z. Bot., 13 (1921) 193.
48 O. Warburg, Biochem., 100 (1919) 230.
49 A.C. Chibnall, Protein Metabolism in the Plant, New Haven, 1939.
50 A.C. Chibnall, Ann. Rev. Biochem., 35 (1966) 1.
51 L.N. Vauquelin and J.J. Robiquet, Ann. Chim. Phys., 57 (1806) 88.
52 A.A. Plisson, J. Pharm., 14 (1828) 177.
53 T.J. Pelouze, Ann. Chim. Phys., 5 (1833) 283.
54 A.W.H. Kolbe, Ann. Chem., 121 (1862) 232.
55 A. Piutti, Gaz. Chem. Ital., 17 (1887) 519.
56 R. Piria, Compt. Rend., 19 (1844) 575.
57 R. Piria, Ann. Chim. Phys., (3), 22 (1848) 60.
58 V. Dessaignes and Chautard, J. Pharm. Chim., (3) 13 (1848) 246.
59 L. Pasteur, Ann. Chim. Phys., 82 (1852) 324.
60 W. Pfeffer, Jahrb. Wiss. Bot., 8 (1872) 429.
61 J.B. Boussingault, Agronomie, Chimie Agricole et Physiologie, vol. 4, Paris, 1868, p. 245.
62 H. Ritthausen, J. Prakt. Chem., 107 (1869) 218.
63 J.L. Proust, Ann. Chim. Phys., (2) 10 (1819) 29.
64 H. Braconnot, Ann. Chim. Phys., (2) 13 (1820) 113.
65 J. Liebig, Ann. Chem., 57 (1846) 127.
66 F. Bopp, Ann. Chem., 69 (1849) 16.
67 E. Erlenmeyer and A. Lipp, Ann. Chem., 219 (1883) 161.
68 E. Cramer, J. Prakt. Chem., 96 (1865) 76.
69 E. Fischer and H. Leuchs, Ber. Dtsch. Chem. Ges., 35 (1902) 3787.
70 H. Ritthausen, J. Prakt. Chem., 99 (1866) 454.
71 H. Ritthausen, J. Prakt. Chem., 103 (1868) 233.
72 H. Ritthausen, Die Eiweisskörper, Bonn, 1872.
73 O. Nasse, Arch. Ges. Physiol., 6 (1872) 589.
74 O. Nasse, Arch. Ges. Physiol., 7 (1873) 139.
75 O. Nasse, Arch. Ges. Physiol., 8 (1874) 381.
76 H. Hlasiwetz and J. Habermann, Ann. Chem., 169 (1873) 150.
77 E. Schulze and E. Bosshard, Landwirtsch. Vers. St., 29 (1883) 295.
78 M. Damodaran, G. Jaarback and A.C. Chibnall, Biochem. J., 26 (1932) 1704.
79 M. Bergmann, L. Zervas and L. Salzmann, Ber. Dtsch. Chem. Ges., 66 (1933) 1288.
80 M. Damodaran, Biochem. J., 26 (1932) 235.
81 G. Salomon, Ber. Dtsch. Chem. Ges., 11 (1878) 574.
82 E. von Gorup-Besanez, Ann. Chem., 98 (1856) 1.

83 P. Schützenberger, Ann. Chim. Phys, (5) 16 (1879) 289.
84 E. von Gorup-Besanez, Ber. Dtsch. Chem. Ges., 7 (1874) 146.
85 E. von Gorup-Besanez, Ber. Dtsch. Chem. Ges., 7 (1874) 569.
86 E. von Gorup-Besanez, Ber. Dtsch. Chem. Ges., 7 (1874) 1478.
87 E. von Gorup-Besanez, Ber. Dtsch. Chem. Ges., 10 (1877) 780.
88 E. Schulze, Landwirtsch. Jahrb., 7 (1878) 411.
89 E. Schulze and J. Barbieri, Ber. Dtsch. Chem. Ges., 10 (1877) 199.
90 E. Schulze and J. Barbieri, Ber. Dtsch. Chem. Ges., 12 (1879) 1924.
91 E. Schulze and J. Barbieri, Ber. Dtsch. Chem. Ges., 14 (1881) 1785.
92 E. Erlenmeyer and A. Lipp, Ber. Dtsch. Chem. Ges., 15 (1882) 1006.
93 E. Schulze and J. Barbieri, J. Prakt. Chem., 27 (1883) 337.
94 E. Schulze and Umlauft, Landwirtsch. Jahrb., 5 (1876) 819.
95 E. Baumann, Arch. Ges. Physiol., 29 (1882) 419.
96 O. Löw, Arch. Ges. Physiol., 22 (1875) 503.
97 O. Löw, J. Prakt. Chem., 31 (1885) 129.
98 W. Pfeffer, Handbuch der Pflanzenphysiologie, Leipzig, 1880.
99 H. Hlasiwetz and J. Habermann, Ann. Chem., 159 (1871) 304.
100 E. Schulze, Landwirtsch Jabrb., 14 (1885) 713.
101 E. Schulze, Z. Physiol. Chem., 47 (1906) 507.
102 D.N. Prianishnikov, Landwirtsch Vers. Stn., 52 (1899) 137.
103 A. Stieger, Z. Physiol. Chem., 86 (1913) 245.
104 Wl. Butkewitsch, Biochem. Z., 16 (1909) 411.
105 H.B. Vickery and C.L.A. Schmid, Chem. Rev., 9 (1931) 169.
106 A. Meister, Biochemistry of the Amino Acids, New York, 1957.
107 J.S. Fruton, Molecules and Life. Historical Essays on the Interplay of Chemistry and Biology, New York, 1972.
108 A. Strecker, Ann. Chem., 76 (1850) 27.
109 T. Weyl, Ber. Dtsch. Chem. Ges., 21 (1888) 1407.
110 E. Drechsel, J. Prakt. Chem., 39 (1889) 425.
111 A. Kossel, Z. Physiol. Chem., 22 (1896) 176.
112 S.G. Hedin, Z. Physiol. Chem., 20 (1895) 186.
113 W.H. Wollaston, Ann. Chim. Phys., 76 (1810) 21.
114 K.A.H. Mörner, Z. Physiol. Chem., 28 (1899) 595.
115 E. Baumann, Z. Physiol. Chem., 8 (1884) 299.
116 R. Willstaether, Ber. Dtsch. Chem. Ges., 33 (1900) 1160.
117 E. Fischer, Z. Physiol. Chem., 33 (1901) 151.
118 E. Fischer, Ber. Dtsch. Chem. Ges., 35 (1902) 2660.
119 F.G. Hopkins and S.W. Cole, Proc. Roy. Soc., 66 (1901) 21.
120 A. Adamkiewicz, Arch. Ges. Physiol., 9 (1874) 156.
121 F. Erlich, Ber. Dtsch. Chem. Ges., 40 (1907) 2538.
122 F. Ehrlich, Ber. Dtsch. Chem. Ges., 37 (1904) 1809.
123 J.H. Mueller, Proc. Soc. Exp. Biol. Med., 19 (1922) 161.
124 W.C. Rose, R.H., McCoy, C.E. Meyer, H.E. Carter, M. Womack and E.T. Mertz, J. Biol. Chem., 109 (1935) 77.

Chapter 46

Biogenetic Hypotheses Derived of the Known Behaviour of Plant Constituents

1. Structural formulas of organic compounds

As we have seen in Chapter 17, it was in the 1850's that the protoplasm theory was first challenged by a group of enzymologists, but it was in the 1890's, after the discovery of cell-free fermentation, that the discipline of biochemistry began to be organized around a chemical program along which the pathways of catabolism were successfully unravelled on the basis of the chemical knowledge of the metabolites involved and of the enzymes concerned. This period of biochemical progress expanded during the years of the present century before World War II. In the same period, most essential aspects of catabolic pathways were clarified, but the anabolic pathways remained unexplained. In this domain, the period considered here is characterized by activities situated at the fringe of organic chemistry and which consisted in intellectual constructions based, when available, on the formation contained in the structural formulas of organic compounds. The unravelling of the structure of such compounds took a long time. There was no special intellectual motivation that oriented the organic chemists towards a study of compounds that were later revealed as of interest for the study of biosynthetic pathways. Organic synthesis was rapidly oriented predominantly towards the creation of innumerable artificial compounds: dyes, oils and perfumes, drugs, etc.

On the other hand, many organic chemists, as Liebig and Dumas had done at the start, were prone to believe that, when they had accomplished the synthesis of an organic compound in vitro, the same pathway was followed in biosynthesis. The first approach, as practised by Liebig, Dumas or Pelouze, was based on chemical statics, involving empirical formulas, and only later was it possible for organic chemists to guess about biosynthetic pathways, and they had to limit their contribution to such aspects until relevant methods of approach became available after World War II.

We should be guilty of "presentist" sin should we fault the organic chemists for the formulation of erroneous reaction sequences. As Claude Bernard and others have repeatedly stated, the organic chemists were basing their conceptions on facts observed in vitro. It was their method to approach the problem statically on the basis of balance sheets and accumulated chemical knowledge. While recognizing their models as hypothetical, they claimed and maintained the chemical nature of anabolism. On the other hand, as we have seen, Claude Bernard, refuting this methodology, was led to the sterile conception of "organic creation" based on physiological but irrelevant facts. On the contrary, Dumas, Liebig and their followers, though short-sighted, remained unwarped by the protoplasm myth and, as was forcefully expressed by Boussingault [1], they kept alive the concept of biogenetic aspects as chemical phenomena.

For Dumas, the anabolic processes in animals resulted from conversions by small steps in oxidation. For instance, he considered that waxes could be converted into fatty acids by such actions.

Liebig, who considered fat biosynthesis, though unnatural, as taking place in animals, believed that sugar could provide fat by removal of oxygen. Dumas, Boussingault and Payen proposed, in 1843, as was recalled above, a chemical pathway for the conversion of sugar into fatty acids when they considered sugar as composed of CO, water and ethylene. Since ethylene can be converted into various alcohols yielding fatty acids, there is no objection "chemically speaking" against a possible conversion of sugar into fatty acids. From the point of view of epistemology, we recognize here the beginnings of a methodology in which a concept of biosynthesis will not have to contradict the accumulated knowledge stated by organic chemists in structural formulas. We have mentioned, in Chapter 42, the equation proposed by Pelouze and Gélis in 1843 for the conversion of glucose into butyric acid. The kind of graphology finally adopted by organic chemists was one of the conditions for a development of studies related to biosynthesis. On the other hand, Bernard did not propose any hypothetical pathway of biosynthesis. He did not believe in such chemical pathways.

2. Berthelot and organic synthesis

We may try to reconstitute the intellectual climate in which the first successes in organic synthesis were accomplished, exerting an effect not only on the scientific community but also in the realm of general ideas. We have emphasized, in Chapter 12, p. 253, the importance of Wöhler's synthesis of urea in its implication of a program of organic synthesis. One leader in this field, Berthelot [2], had explained his ideas on scientific philosophy in a letter to his friend Renan, dated November 1863. Berthelot presents himself as a member of the positivist school, founded by Auguste Comte:

"Positive science, he writes, does not look after the original causes or after the final causes of things; but it proceeds by establishing facts and by linking them one to another, by immediate relations". (Translated by the author.)

Positive science establishes chains of causal relations. It is in this context that we must understand the statement of Berthelot about the necessity of banning life from all explanations concerning organic chemistry (Chapter 12, p. 261). There is no place for vital force in a positivist philosophy. In 1859, Berthelot [2] had published in the *Revue Germanique* an article in which he presented the ideas he was going to develop, in 1860, in his book *Chimie Organique Fondée sur la Synthèse* [3]. He first recalls that, since the time of Lavoisier, it has been possible to decompose any inorganic compound and inversely to reconstitute the primitive compound by the association of its components, a reconstitution that for a long time remained impossible for organic compounds. The impotence of chemistry in its attempts to reproduce the association of carbon and hydrogen and the various compounds resulting from such association has long been considered as the expression of an impassable chasm between inorganic chemistry and organic chemistry. Between 1830 and 1860, a remarkable succession of enquiries provided, on the basis of mild decomposition, a large number of derivatives that have contributed to the analytical foundation of a classification of organic substances.

"But it was impossible to go up the ladder, starting from the chemical elements in order to form, through the participation of the affinities we used to rely upon in inorganic nature, such compounds as hydrocarbons, and afterwards alcohols and substances of higher complexity" (p. 47). (Translation by the author.)

Berthelot (slighting Wöhler's synthesis of urea accomplished in 1828) states that until 1860 all experimenters who have written on organic chemistry started from the "proximate principles" obtained from organisms.

"In general they have proceeded from the ligneous or from starch to sugar, from sugar to alcohol and finally from alcohol to hydrocarbons; in short they have started from the simplest compounds among those found in organisms and they have proceeded downward by successive analysis, through a sequence of simpler and simpler realities, until reaching the binary combinations and the elements. This was a curious, though necessary, mixture of chemistry and natural history, which deprived science of a part of its abstract rigor". (Translation by the author.)

Berthelot, pointing to the incapacity of producing organic compounds calls attention to a possible confusion between two categories of problems. The chemist will never pretend to produce a leaf or a muscle in the laboratory: these problems belong to what was designated as physiology at the time. But if such structures escape the field of the activity of the chemist, the formation of the "proximate principles" belongs to the field of organic synthesis. Berthelot believed he had established the principles of the chemical synthesis of a great variety of organic compounds and that, as a consequence, a large body of research could proceed from the known to the unknown without leaning on ideas other than those resulting from a purely chemical and physical study.

"Instead of originating in life phenomena, organic chemistry now possesses an independent basis; it is now its turn of rendering to physiology the help it has for such a long time received from it" (pp. 52—53).

The basis of the development was recognized by Berthelot in his synthesis of acetylene from carbon and hydrogen, associated under the influence of electricity. From acetylene he derived a number of hydrocarbons. He also obtained such compounds in other ways, particularly by starting from the simplest binary compounds, such as carbon dioxide, carbon monoxide and water, a method that presents the particular interest of

"starting from the same origin as living nature, though through quite different contrivances. For, it is from water and carbonic acid that plants and animals form the various principles composing their tissues" (p. 54).

The formation of an increasing number of organic compounds by chemical synthesis leads to a capital demonstration:

"Indeed, by the fact of the formation of organic compounds and by an imitation of the processes which govern their formation in plants and animals, it can be established that the chemical effects of life are a consequence of the display of ordinary chemical forces, as well as the physical and mechanical effects of life result of a display of purely physical and mechanical forces. In both cases the molecular forces implied are identical, for they produce the same effects" (pp. 58—59).

Not only did Berthelot recognize the value of a study of the formation of organic compounds, and of the causes determining their formation, as providing a chemical interpretation of vital phenomena, but he also recognized that the studies on organic synthesis led to a deeper knowledge of molecular forces and of the laws related to the action of such forces. The knowledge of these laws leads to a particularly fruitful extension: the power of forming innumerable other organic compounds. Not only had, for instance, the synthesis of neutral fats allowed Berthelot to obtain the natural fats known at the time, and to prove that fats were neutral glyceric salts of the fatty acids, but it led to the synthesis of a large number of unnatural fats.

Berthelot's book, *Chimie Organique Fondée sur la Synthèse* [3] (1860) brought to a culmination the path opened bt Wöhler's synthesis of urea in 1828. The book established Berthelot's reputation, and he was appointed professor at the Collège de France, where he delivered his introductory lecture on February 2, 1864 [2]. In that lecture [2], Berthelot, after recalling the history of the development of organic chemistry, referred to the interplay of physiology (meaning what we call biology *) and of organic chemistry.

"A truth is always fertile, every development of general concepts is at the origin of an infinity of consequences in the different theoretical sciences,

* The term biology, designating the science of life, was introduced independently by Lamarck and by Treviranus. While it was immediately adopted by German scientists, it took longer in France to substitute it for physiology. It was De Blainville who introduced it. Aug. Comte defines it as encompassing the anatomy and the physiology of all kinds of animals and plants. (On this point see Mendelsohn [4] and Holmes [5].)

References p. 192

and applied sciences as well. Considering other sciences, it will suffice to mention physiology: those who cultivate this science know what light is projected on it by organic chemistry, and which level the interplay of both sciences may reach. The general problems of nutrition in living organisms are chemical problems; such are also those of respiration. The progress, in the domain of such problems, is based on data obtained by organic chemistry".

Boussingault, as stated above, had forcefully expressed the view of the Dumas school of the chemical nature of life. Berthelot accomplished this role of champion for Liebig's school. He had started his work on organic synthesis in the laboratory of Pelouze. Pelouze, while he was professor at the Ecole Polytechnique where he had replaced Dumas, had served as *suppléant* to Thenard at the Collège de France and he succeeded him in that chair. Pelouze was the closest friend of Liebig among Parisian chemists. As stated by Holmes [5]:

"Pelouze... had become allied with Liebig and Berzelius during the controversies between them and Dumas over the composition of organic compounds. In 1836 Pelouze had become a kind of unofficial representative in Paris of Liebig's chemical school" (p. 77).

When he mentioned, in his book of 1860, the precursors of his own work, Berthelot, among others, referred to Gay-Lussac (the master of Liebig), Wöhler and Liebig.

Berthelot's "chemicalism" took the position of a public scandal when he opened the preface of his book *Les Origines de l'Alchimie* [6] with the sentence "The world is without mystery today" which was immediately used as a stone to lapidate its author as a smug scientist, even a sciolist. Whereas the sentence had been written in the context of a comparison between alchemy and modern science, it was considered, by the litterati of the time, as the motto of an aggressive scientism. This episode has been most aptly told by Virtanen [7]. It led to a proclamation of the "bankruptcy of science" and to a breaking out of manifestations of antiscientism *. In spite of such attacks, Berthelot never lost his faith in progress through science, and he fortunately did not live long enough to see the First World War. As a chemist, he won, until his

* These controversies are echoed in a number of literary works, in which Berthelot inspires the features of a character, such as Bertheroy in Zola's *Paris* or Jacques Dubardeau in Giraudoux' *Bella*. (See Virtanen [7].)

death in 1907, a universal recognition based on an impressive list of contributions to organic chemistry (see Crosland [8]).

To Berthelot's fame as a chemist, was added his role in the political life of the French Third Republic, and the position of his family in public and intellectual life. His legend until the turn of the century competed with those of Pasteur and of Bernard. Today he is mainly spoken of as an example of scientism, and also of sterile conservatism. The latter aspect appeared in his reluctance to accept the atomic theory. Like Sainte-Claire-Deville, he refused to grant any reality to the atom. These two on one side were violently opposed, in this issue, to Wurtz on the other side; and the controversy reached a climax during a meeting of the Académie des Sciences in 1877. As a state minister, Berthelot prevented the teaching of the atomic theory in schools for a long time. The biologist Caullery [9] has recalled that as late as 1887 the teachers at the Ecole Normale Supérieure were wont to refer sarcastically to the atomic theory. As stated by Taton [10], in reference to Dumas and Berthelot who were both state ministers,

"Experience shows that it is always dangerous to confer too much power of criticism upon even the most eminent scientists, for there are some who, with age, turn theories into unassailable dogma against which they allow no criticism. And if their powers are too wide, some of them may reduce their young adversaries to utter silence and thus brake the progress of science. Jean-Baptiste Dumas and Marcelin Berthelot were two eminent scientists who for a time enforced a scientific dogmatism against which it was very difficult to struggle". (Translation by Pomerans [10], p. 148.)

Persistently sticking to the doctrine of equivalentism and short-sightedly opposed to the development of the concept of the structural formula, Berthelot paradoxically supported obstacles to the development of a principle he had formulated himself.

3. Importance of the concept of the structural formula

While such a reaction as suggested by Pelouze for the biosynthesis of fatty acids from sugar was still situated in the frame of studies based on the empirical formula, the progress of organic chemistry led to a formulation of possible mechanisms for biosynthesis. Balance experiments performed by Lawes and Gilbert

References p. 192

Plate 155. Charles Adolphe Wurtz.

[11] afforded decisive data on the conversion; these authors also demonstrated that fats could arise from proteins. In 1878, Nencki [12] proposed a scheme of chemical reactions according to which lactic acid derived from glucose could be degraded to acetaldehyde, two molecules of which condensed to form butyric acid. The development of the concept of structural formulas (see Chapter 12) incorporated into a graphology the scientific laws ruling the formation and the metamorphosis of organic compounds, the structural formulas being intended to show the relations between organic compounds in the course of their transformations. As stated by Hempel [13] (p. 249), the conditions of scientific explanation command a logical deduction leading to a description of the empirical phenomena to be explained (in biosynthetic studies, the *explanandum* is the production in vivo of a kind of biomolecule). Not only do these aspects consist in statements of antecedent conditions (in this case the intermediary metabolites and the enzymes involved) but also of general scientific laws governing the transformations of organic compounds (implied by the chemist in the structural formula arrived at by way of chemical synthesis). In the frame of genetic epistemology, the *explanans* (the biosynthetic pathway) was based on statements of antecedent conditions, plus general laws. The predictive value of the possibilities implied in the structural formula have therefore afforded a fruitful deductive process in the context of the interplay of chemistry and biology. The transition from the possible pathways to those actually implemented at a definite site of an organism involved a deductive method and an experimental demonstration that could only be accomplished at a later stage.

4. Emil Fischer proposes a new alliance (1907)

The interplay of organic synthesis and biology was called forth by Fischer [14] in 1907. Fischer devoted to this topic the Faraday lecture he delivered at the Royal Institution on October 18, 1907. Fischer recalls that in its early youth organic chemistry was closely connected with biology: the materials investigated by organic chemistry were mostly products of biological origin, and it is in the study of carbohydrates, proteins and vegetable acids that

References p. 192

Plate 156. Emil Fischer.

elementary analysis was elaborated by Lavoisier, Gay-Lussac, Berzelius, Dumas and Liebig.

During the first decade of the nineteenth century, the living world was the source of the isolation of many organic compounds, the list of which had been increasing, reaching several hundreds by the time of Fischer's program-lecture. But, he notes how small is this number compared with the 130 000 carbon compounds known at the same time. Fischer recognized that during the latter half of the 19th century organic chemistry became separated from biology:

"It cannot be mere chance that the most famous of Liebig's pupils, A.W. Hofmann, A. Kekulé and A. Wurtz did not follow the example of their great teacher, whose chief triumphs were won by the use he made of chemical methods in solving biological problems. Perhaps they were restrained by the feeling that, mainly through his influence, physiological chemistry has been developed into a separate discipline, which should be cared for by men who could devote themselves entirely to its service. Such subdivision of labour undoubtedly has many advantages; the disadvantage would have outweighed these had it precluded interchange of experiences and friendly cooperation of workers in the two fields; the history of both sciences, however, affords ample proof that such has not been the case".

Noting that physiologists have availed themselves of the developments of organic chemistry and that, on the other hand, chemists have in many cases been stimulated by the existence of physiological studies, Fischer adds:

"But organic chemistry will certainly never be content to act as a mere handmaid of biology. This is impossible, as the theoretical and technical problems which she is called upon to consider are already numerous and they cannot fail to increase in number and importance in the future. But I do consider it not only possible but desirable, that the close connexion of chemistry with biology which prevailed in the days of Liebig and Dumas should be reestablished, as the great chemical secrets of life are only to be unveiled by co-operative work".

In this lecture, Fischer indicated the part which organic chemistry can play. He referred to the explanation of the formation of *d*-glucose and fructose alone in photosynthesis by his own work on synthesis in the sugar group. But he accepted that data of organic chemistry have so far failed to give a complete explanation of biochemical processes, and greater returns should be sought in other directions.

"The ultimate aim of biochemistry is to gain complete insight into the unending series of changes which attend plant and animal metabolism. To accomplish a task of such magnitude, complete knowledge is required of each individual chemical substance occurring in the cycle of changes and of analytical methods which will permit of its recognition under conditions such as exist in the living organism. As a matter of course, it is the office of organic chemistry, especially of synthetic chemistry, to accumulate this absolutely essential material. The chemical constitution of hundreds of carbon compounds which occur naturally has already been determined and their more important properties have been established. But far more remains to be done".

That far more remained to be done was the topic treated in the body of Fischer's lecture. He pointed to the many uncertainties still present in the knowledge of fat metabolism. He noted that only few polysaccharides had been successfully synthesized, and pointed to the desirability of organic synthesis in the field of dextrins, gums and "similar undeciphered substances". This applied even more to proteins:

"As they are among the most complex substances produced in the living world and are concerned in all the vital activities of the cell, a complete comprehension of their nature must obviously precede the development of biological chemistry".

At this point, Fischer made a survey of the position of protein chemistry in 1907. In this most valuable picture he pointed to the desirability of pursuing the study of synthetic polypeptides and accomplishing the gigantic task of the synthesis of proteins. Enzymes had been shown to depend in their specific action, on the structure and configuration of the object they attack,

"But, unfortunately, we know practically nothing of the composition of the enzymes, as the complete isolation of an enzyme has never been accomplished. From observations hitherto made, it appears in a measure probable that they are derived from proteins and possess a protein-like character. If this be so, it may be hoped that the experience gained with the protein will be of service in the investigation of enzymes".

If organic chemists have prepared alizarin and indigo artificially in huge quantities and if many other colorants have also been industrially synthesized,

"Our ignorance is correspondingly great of most of the blood colouring matters as well as of many coloured constituents of our own bodies: of the hair, the skin and the eyes".

To quote Fischer, concluding his program-lecture calling on organic chemists to devote their activities to problems of importance for biochemistry:

"In fine, the aid of synthetical chemistry is required in every direction in arriving at a clear understanding of structure and of change. The methods at our disposal in the laboratory are doubtless altogether different from those which come into operation in the living world but chemists are already trying to effect changes in carbon compounds by means of so-called mild interactions, under conditions comparable with those which prevail in the living organism. . . . In fact the effort is already being made to co-operate with biology; it is clear that a section of the forces of organic chemistry is being directed once more towards the goal from which it set out. The separation from biology was necessary during the past century while experimental methods and theories were being elaborated; now that our science is provided with a powerful armory of analytical and synthetical weapons, chemists can once more renew the alliance both to its own honour and to the advantage of biology" (p. 1765).

5. Sir Robert Robinson on the synthesis of alkaloids (1917)

An important step was the publication, in 1917, of a paper by Sir Robert Robinson [15] entitled. "A theory of the mechanism of the phytochemical synthesis of certain alkaloids". Referring to phytochemical syntheses, at the start of his paper Robinson states that little progress had been made in ascertaining the nature of these syntheses

"or even in the less ambitious task of formulating possible mechanisms based on laboratory analogies".

"The details of the schemes which have been suggested with but few exceptions, involve reactions for which little or no parallel exists in synthetical organic chemistry under conditions approximating to those obtaining in a plant".

The starting point of Robinson's paper was his realization of a synthesis of the alkaloid tropinone from succinaldehyde, methylamine and acetone in aqueous solution (R. Robinson [16]). Robinson was convinced that such mechanisms as that described by Pictet of a synthesis of nicotine by pyrrogenic reactions of pyrrol derivatives, were calling on the command by plants of powerful reagents that were not present in plants. On the other hand, his tropinone synthesis, taking place under mild conditions

References p. 192

Plate 157. Sir Robert Robinson.

"on account of its simplicity, is probably the method employed by the plant" (p. 877).

The argument of simplicity reflects the current use of Occam's razor in the intellectual constructions of the chemists. Robinson developed a theory giving the role of intermediates, in the biosynthesis of several alkaloids, to such amino acids as ornithine and lysine and to presumed products of carbohydrate breakdown, such as acetone dicarboxylic acid ($HOOC{-}CH_2{-}CO{-}CH_2{-}COOH$). As Fruton [17] states:

"Not only were such theories intellectually satisfying but also they proved to be of considerable practical value in the determination of the structure and in the laboratory synthesis of many alkaloids" (p. 461).

6. Structural relationships between different classes of organic compounds

In 1922 there appeared a review paper by J.A. Hall [18] based on a concept which had been introduced by Kremers [19] in a study on the biosynthesis of the constituents of the volatile oil of peppermint. Kremers proposed that the menthone group of organic compounds found in peppermint (*Mentha piperita*) and the carvone group of compounds present in spearmint (*Mentha spicata*) were products of the reduction of a single precursor, citral. In building such kinds of hypothetical biosynthetic pathways based on molecular structures, Hall recognized limitations imposed by technical difficulties, such as the lack of suitable methods for isolating other than those stable compounds capable of enduring drastic methods.

With respect to unstable intermediary metabolites, Hall [18] states that

"their isolation from the living plant appears to be improbable, and their fixation in plant materials prepared for examination may prove to be impossible".

"The lack of experimental methods has necessitated an approach to the problem along theoretical lines in which cognizance has been taken of such organic chemical experience as could be translated into biochemical thought and due regard given to such biological experience as could be translated into organic chemical thought" (p. 306).

Bernhauer [20] has written a critical review of the theories propounded by organic chemists to account for the relationship of

References p. 192

sugars and their derivatives to several classes of naturally occurring substances.

On the other hand, isoprene polymerization had been proposed by several organic chemists (review by Bogert [21]) to explain the formation of terpenes, polyene pigments, sterols, caoutchouc, etc. *. It is important to point to a concept introduced by Schöpf [23] who stated that, while it is difficult to draw broad conclusions as to precursors from the constitution of an individual substance, if a large number of compounds of related constitution are known it is possible to develop a kind of "comparative anatomy" from which conclusions can be drawn about the precursors. He nevertheless astutely pointed to the notion that substances which may appear to be biogenetically related may be derived from a common precursor by different routes. We see here the first formulation, by an organic chemist, of a biochemical principle that was formulated much later (see volume 29A, Chapter 1) by comparative biochemists, the co-linearity of biosynthesis and molecular evolution and the necessary knowledge to establish this co-linearity — knowledge of the pathway of reactions leading to the compound concerned, and of the enzymes involved. Emde [24] also pointed to structural regularities in plant products and formulated a "reconstructive biochemistry" on the basis of these regularities.

Hall [18] applied a similar reconstructive reasoning to the question of the biogenic origin of quinic acid. From a review of the examples of such reconstructive schemes extant at the time of his writing, Hall concluded that, if the problems of biosynthesis cannot be solved by mere formulations of structure, these formulations can suggest lines of laboratory approach, but that such approach must be radically different from conventional organic approach.

7. Retrospect

Used to the interpretation of chemical synthesis as a way of reaching knowledge of molecular structure, the organic chemists, facing

* As we shall see later, the "isoprene rule" became firmly established by Ruzicka after 1920 (see Ruzicka [22]).

the growing collection of "immediate principles" obtained by analyzing organisms, were bound to annex this field to their methodological domain. They synthesized a number of these compounds, including substances of interest to biochemists. From the knowledge they obtained on natural compounds, the organic chemists recognized that a degree of order could be attained if the diversity of the natural compounds was interpreted in terms of biogenesis. By about 1950, the bulk of natural substances was classified into a number of main biogenic categories, but

"still only in terms of hypothetical transformations, which convinced chemists wholly, and biologists not at all". (Bu'Lock [25], p. 2.)

As Fruton [17] states (p. 470):

"During the period 1920—1940, some organic chemists considered such "biogenetic hypotheses" to reflect the chemical processes occurring in the plant, and sought to effect the artificial synthesis of alkaloids under 'physiological conditions'." But, after World War II it was generally agreed that

"... the comparison of structures per se gives no information about the details of the mechanisms of biosynthetic reactions. It is the task for the biochemist to determine these by appropriate experiments." (Robinson [26], p. 1.)

Appropriate experiments became feasible when the biochemists had at their disposal a number of new methods: isotopic tracer techniques; use of mutants; isolation of pure enzymes, etc. These methods, together with the treasure of knowledge provided, permitted an acceleration of the unravelling of biosynthetic pathways, which will be described in Part V. But before this acceleration of biochemical history took place, attempts had been made towards the solution of biosynthetic problems, in plants as well as in animals. These first biochemical approaches are related in the remaining Chapters of the present volume.

References p. 192

REFERENCES

1 J.B. Boussingault, Ann. Chim. Phys., 14 (1945) 419.
2 M. Berthelot, Science et Philosophie, Paris, 1886.
3 M. Berthelot, Chimie Organique Fondée sur la Synthèse, Paris, 1860.
4 E. Mendelsohn, Hist. of Sci., 3 (1964) 40.
5 F.L. Holmes, Claude Bernard and Animal Chemistry, Cambridge, Mass., 1974.
6 M. Berthelot, Les Origines de l'Alchimie, Paris, 1885.
7 R. Virtanen, Marcelin Berthelot. A study of a scientist's public role, Lincoln, 1965.
8 M.P. Crosland, in C. Gillispie (Ed.), Dict. Sci. Biogr., vol. 2, New York, 1970, p. 63.
9 M. Caullery, La Science Française depuis le 17e Siècle, Paris, 1948.
10 R. Taton, Causalité et Accidents de la Découverte Scientifique, Paris, 1955. (English translation by A.J. Pomerans: Reason and Chance in Scientific Discovery, New York, 1957)
11 B.J. Lawes and J.H. Gilbert, Phil. Mag., (4) 32 (1866) 439.
12 M. Nencki, J. Prakt. Chem., 14 (1878) 105.
13 C.G. Hempel, Aspects of Scientific Explanation and Other Essays in the Philosophy of Science, New York, 1965.
14 E. Fischer, J. Chem. Soc., 91 (1907) 1749.
15 R. Robinson, J. Chem. Soc., 111 (1917) 876.
16 R. Robinson, J. Chem. Soc., 111 (1917) 762.
17 J. Fruton, Molecules and Life: Historical Essays on the Interplay of Chemistry and Biology, New York, 1972.
18 J.A. Hall, Chem. Rev., 20 (1922) 305.
19 R.E. Kremers, J. Biol. Chem., 50 (1922) 31.
20 K. Bernhauer, Grundzüge der Chemie und Biochimie der Zuckerarten, Pt. 3, Berlin, 1933.
21 M. Bogert, Chem. Rev., 10 (1932) 265.
22 L. Ruzicka, Proc. Chem. Soc., (1959) 341.
23 C. Schöpf, Ninth Int. Cong. Pure and Applied Chemistry, vol. 5, 1934.
24 H. Emde, Helv. Chim. Acta, 14 (1931) 881.
25 J.D. Bu'Lock, The Biosynthesis of Natural Products, London, 1965.
26 R. Robinson, The Structural Relations of Natural Products, Oxford, 1955.

Chapter 47

Ureotelism and Uricotelism

1. Uric acid

For a long time before biochemistry had its own institutions in academic departments, biochemical studies were accomplished in the departments of chemistry of the medical faculties mainly as a consequence of medical motivations. Not only Berthollet, but also Fourcroy, whose personality was already mentioned in Chapter 5, started in medicine, and he may be considered as a pioneer of "animal chemistry". Fourcroy belonged to that category of mind which is prone to be led by the solicitations of actuality, whether they be the opening of a large cemetery, a fog, a falling of stones from the sky or the re-shaping of a higher education system. Smeaton has compiled a long list of Fourcroy's publications giving an idea of the wide breadth of his interests, in the pursuit of which he was seconded by his colleague Vauquelin, "a skillful and indefatigable experimental chemist" (Smeaton [1], p. XIX). Vauquelin was a pharmacist who had served as an apprentice in the pharmacy of a cousin of Fourcroy who helped him in his course of studies and took him as his assistant and demonstrator. Later, Vauquelin became manager of a pharmacy, and Fourcroy's two sisters looked after him until his death. Vauquelin became professor at the Collège de France in 1801. He resigned this post in 1804 to become professor at the Museum d'Histoire Naturelle and director of the School of Pharmacy. Besides his collaboration with Fourcroy, Vauquelin accomplished important studies on the chemistry of minerals.

We have, in Chapter 5, referred to the work of Fourcroy and the introduction of the concept of the "principes immédiats" (proximate principles, immediate principles). Fourcroy's insati-

Plate 158. Louis Nicolas Vauquelin.

able curiosity and versatility was catalysed by his conviction of the importance of chemical knowledge for the benefit of medicine. He started as an anatomist, but he soon turned to chemical studies and lost no opportunity of using animal substances for analysis. At the time, before the progress accomplished by Gay-Lussac and by his disciple Liebig, analytical methods were insufficient for isolating pure compounds from animal materials. One of Fourcroy's ever-present preoccupations was the determination of constants expressed by numbers, such as melting point or solubility. Such measurements had seldom been performed during the 18th century.

After the recognition, by Lavoisier, of the constitution of water, Fourcroy became the greatest opponent of phlogistonic theories and an outstanding lecturer and author of scientific treatises. His work on the compounds of magnesium and mercury were valuable contributions to the new chemistry. His work on animal chemistry, though bearing the mark of still unsatisfactory analytical procedures, reveals the deep interest of a pioneer. While still a medical student, Fourcroy had translated from the Italian the classical treatise of Ramazzini on professional diseases. This translation he accomplished at the suggestion of the Société Royale de Médecine which started as a commission primarily appointed for the study of epidemics. It developed later into a sort of academy. The Société Royale de Médecine, of which Fourcroy had become a member, owned a collection of gallstones and urinary stones. We recalled above the attention paid by Black to urinary calculi and the importance of this interest in the origin of his significant chemical and biochemical studies on carbonic anhydride. The importance of chemical studies on "stones" has been emphasized in an excellent review by Coley [**2**].

From urinary calculi, Scheele [3] had, in 1776, isolated a new acid principle, sparingly soluble in water, giving no reaction with dilute mineral acids, but dissolving rapidly in alkalis. When treated with concentrated nitric acid, this material became pink or purple (later called "the murexide reaction"). Scheele constantly found, in human urine, this compound first isolated by him from urinary calculi. Scheele, who accomplished his work in the course of the diverse forms of his activities as a pharmacist, had been a student of Torbern Bergman at Uppsala, and his former master added a supplement [4] to the essay of Scheele, confirming his discovery

References p. 207

Plate 159. Carl Wilhelm Scheele.

of a new acid principle in urinary stones and in urine. This principle was called "lithic acid" by Guyton de Morveau [5] in 1786.

Stimulated by the collection of stones of the Société Royale de Médecine, Fourcroy [6] began a comparative study of these pathological concretions. He concluded that human urinary calculi mainly consisted of "lithic acid" although cavities in the stone often contained phosphates of sodium and ammonium. Fourcroy was surprised to find no "lithic acid" in a renal calculus of a horse, which he recognized as composed of calcium carbonate and phosphate. Was lithic acid, asked Fourcroy, only present in man? In the same paper, Fourcroy showed that, on distillation, lithic acid yielded much nitrogen.

At the same time, in England, Wollaston [7] had started his experiments on the action of acids and alkalis on urinary calculi. Wollaston (1766—1828) had taken his M.D. degree in 1793 and he had practised medicine for some time. It was during this period that he accomplished his studies on calculi and that he made the capital observation on the presence of "lithic acid" in the chalky deposits of gout, to which we shall return.

In 1800, Wollaston abandoned medicine and embarked on metallurgical research which made him famous. For instance, he separated palladium from platinum. Fellow of the Royal Society since 1794, he was awarded the high distinction of the Copley Medal in 1802. (On Wollaston, see Peters [8].)

Another pathologist, Pearson, who had been a pupil of Black and who later became senior physican at St. George's Hospital, London, was interested in the composition of human urinary calculi. He studied more than three hundred of these, digesting them with sodium hydroxide solutions. According to him [9], the main constituent was not an acid as Scheele had proposed, but an "animal oxide" for which he suggested the name "ouric oxide". Pearson's publication revived the interest of Fourcroy [10] in "lithic acid". The "ouric oxide" of Pearson, he ascertained, when moistened, reacted as an acid. Fourcroy proposed to demise the terms "lithic acid" and "ouric oxide" and to adopt the name "uric acid", a suggestion that won general recognition.

In the paper just mentioned, Fourcroy invited his readers to send calculi to him, together with clinical information on age, state of health and other data concerning the patients. This appeal resulted in a collection of a large number of specimens. An analy-

References p. 207

Plate 160. William Hyde Wollaston.

sis by Fourcroy and Vauquelin [11] of the first hundred received was reported in a preliminary account of 1798, and a more detailed account followed after 300 calculi had been analyzed.

In these studies, Fourcroy and Vauquelin were motivated by therapeutical considerations: the hope of devising a method for dissolving the calculi within the patient's organism. They found in about 30 per cent of the calculi analyzed, in addition to uric acid and phosphate of lime, phosphates of ammonia and magnesia. They also detected oxalate of lime in some.

2. Urea

The compound which now bears the name of urea was first isolated from urine by Boerhaave, who called it "native salt of urine" ("*Sal nativus urinae*"). (Literature in Drabkin [12], p. 202.) F.M. Rouelle (Rouelle junior) isolated it in an impure state in 1773 [13]. He extracted with alcohol the residue left by the evaporation of urine and obtained what he called a "matière savonneuse", rich in nitrogen, giving by analysis more than half its weight of "volatile alkali", and giving on fermentation carbonic acid and "volatile alkali". (On the history of the organic chemistry of urea, see Werner [14], and Kurzer and Sanderson [15].)

Urea was obtained in a crystalline form by Fourcroy and Vauquelin [16]. Preoccupied with the pathological problem of the formation of urinary calculi, Fourcroy and Vauquelin turned their attention towards the chemistry of urine, the matrix of stone formation.

In 1797 they read before the Académie des Sciences a paper, on the urine of horse compared with that of man, which was published in 1797. This important contribution was not, at the time of its reading, appreciated as much as it deserved. A journalist reporting on it in the official French newspaper "Le Moniteur" considered the subject as trite and "singularly unfortunate" (Crosland [17], p. 154). In horse's urine, the usual reagents showed the presence of what we now call carbonates of calcium and of sodium, chlorides of sodium and potassium and sometimes some sulphate. The residue left by evaporation was partly soluble in alcohol. On evaporation, the alcoholic solution first gave potassium chloride. On evaporation to dryness, a mass of white crys-

References p. 207

tals was obtained. Partly redissolved in water, it gave, on addition of hydrochloric acid, sodium chloride and benzoic acid. The authors concluded that sodium benzoate was detected (also in cow's urine). They interpreted this finding to be a result of the vegetarian diet. We know now that Fourcroy and Vauquelin had obtained a decomposition product of hippuric acid (benzoyl glycine). Hippuric acid was later isolated and named by Liebig [18], in 1829.

Of the material extracted by alcohol from the solid residue of evaporated urine, there remained, after the removal of what Fourcroy and Vauquelin obtained as sodium benzoate, a portion that, on the addition of nitric acid to the aqueous solution, gave a crystalline precipitate, which turned yellow and finally red.

Commenting on these experiments, Smeaton [1] remarked that the nitrate of urea is white and that the coloured precipitate obtained by Fourcroy and Vauquelin was coloured because it was precipitated from a very concentrated solution containing impurities, and was not washed. When alkali was added to the precipitate, a red oil was produced, giving a white solid on cooling. While behaving like an alkali by giving a salt with nitric acid, the substance was not like the known alkalis. It melted on heating, gave ammonia and left no residue. Smeaton [1] notes that, a short time before, W. Cruickshank (in Rollo [19]) obtained "shiny scaly substances resembling acid of borax" by adding nitric acid to evaporated urine. He did not pursue the study of the compound.

In the same paper, Fourcroy and Vauquelin [16] considered the putrefaction of horses' urine and suggested that the formation of calculi was the result of a "fermentation" in the body, and of a liberation of ammonia making the urine alkaline, while fresh urine was slightly acid. The alkalinization resulted, they thought, in a precipitation of calcium phosphate and of magnesium and ammonium phosphates.

In a second paper, also published in 1799, Fourcroy and Vauquelin [20] defined the compound they christened "urea" and gave a detailed description of its properties. To prepare urea, they evaporated urine to dryness, extracted the residue with alcohol and obtained the crystallized compound from the alcoholic solution. The crystalline product, which was still impure, they described as deliquescent, yellowish and presenting a characteristic odour. By the addition of nitric acid to its solution it was precipitated in the form of white crystals. By distillation of its aqueous solutions it gave ammonia, which the authors had already detected as putrefaction products of urea. The high nitrogen content of urea led Fourcroy and Vauquelin to formulate a physiological theory according to which the function of the kidney consisted in the removal of nitrogen from the body in the form of urea, just as the lungs were considered as removing carbon in the form of carbonic anhydride and the liver as removing hydrogen in the form of the fatty portion of bile.

In 1808, Fourcroy and Vauquelin [21] described purer preparations of urea. The method of preparation started by evaporation, not to dryness, but to a syrupy consistency, and nitric acid was added. After the mixture had been cooled for several hours, the precipitated white crystals of nitrate of urea were washed, filtered and dried between papers. The crystals were dissolved in water, and potassium carbonate was added until the nitric acid was neutralized. Evaporation to dryness followed, and the residue was extracted with alcohol which dissolved the urea and left a residue of potassium nitrate. By evaporation of the alcoholic solution of urea, white crystals were obtained. On this basis, Fourcroy and Vauquelin retracted their first conclusion that the colour of urine was due to urea.

In the same paper, Fourcroy and Vauquelin described the action of heat on pure urea crystals. We know that by heating urea, cyanuric acid and biuret are obtained, besides other products. On this point Fourcroy and Vauquelin made a mistake, as they stated that their product was uric acid. From this erroneous observation they derived a new pathological theory, according to which the uric acid calculi were formed in the bladder by the same reaction as the one they believed they had reproduced by heating urea. These concepts led them to study the action of heat on uric acid. As we know, this action leads to urea, cyanuric acid, hydrogen cyanide, ammonia and other products. Having observed the formation of urea, Fourcroy and Vauquelin reached the erroneous conclusion that there was a close chemical relation between urea and uric acid and that each of them could be converted into the other in metabolism, as can happen on heating. This concept was later at the origin of many erroneous theories.

In 1820, urea was obtained in a pure state by J.L. Proust [22]. He decomposed urea nitrate with carbonate of lead and obtained

References p. 207

almost colourless crystals. The correct empirical formula of urea was established in 1824 by W. Prout in the course of a study of some of the immediate principles of urine. (See Werner [14].) Wöhler, in 1828, synthesized urea by combining ammonia with cyanic acid. (See Chapter 12.)

It was suspected for some time that cyanate of ammonia and urea might be one and the same substance. In 1830 Liebig and Wöhler showed that cyanate of ammonia was different from urea. (See Werner [14].) Though Regnault [23], in 1838, used for the first time the term "carbamide", he failed in identifying urea as being carbamide. Werner retraced the long delay of uncertainties which had preceded the recognition of the carbamide structural formula of urea

$$O{=}C\begin{matrix} \diagup NH_2 \\ \diagdown NH_2 \end{matrix}$$

3. From comparative pathological chemistry to chemical zoology

Fourcroy and Vauquelin [24], in 1804, proposed a classification of calculi. Among urinary stones of mammals, they distinguished those consisting mainly of carbonate of lime, which they recognized as the only constituent in stones from herbivorous animals. Other calculi were mainly composed of phosphate of lime, and yet others of oxalate of lime. Uric acid is a common constituent of human calculi but the authors, as Fourcroy had already done in 1793 about a stone obtained from a horse, expressed their suprise at the lack of uric acid in the calculi of any mammalian species other than man. They suggested that this apparent specificity may have been because they only analyzed calculi of herbivorous animals.

In 1811, after Fourcroy's death, Vauquelin [25], obtained urines of lion and tiger and showed that the urine of these carnivores contained much urea but no uric acid. On the other hand, Fremy and Vauquelin had, around 1805 (see Smeaton [1]), analyzed the excreta of several birds (including an ostrich) and, in spite of their herbivorous habits, found large amounts of uric acid but no urea [26].

In 1822, Vauquelin [27] reported that the excreta of snakes

were constituted of almost pure uric acid. In 1815, Prout [22] had already recognized uric acid as an important constituent of snake's excreta *.

From such comparative data, approached from the domain of chemical zoology, resulted the concept of "ureotelism", i.e. of a predominance of urea as end-point of amino acid metabolism as first observed in mammals. This was opposed to a predominance of uric acid as observed in birds and snakes (Sauropsida). This was already clear-cut systematic distinction, which was confirmed (see below) by experimentation and by more analytical data. For instance it was confirmed that, in the urine of man, urea nitrogen corresponds to 80—85 per cent of total nitrogen, the proportion of uric acid nitrogen amounting to 1.5—4.0 per cent and the ammonia nitrogen to 3—4 per cent (Mattice [29]). In the urine of the frog (*Rana temporaria*) the same pattern obtained: 82 per cent of urea nitrogen, traces of uric acid nitrogen and 5 per cent of ammonia nitrogen (Toda and Taguchi [30]). On the other hand, in the material excreted by the chicken, uric acid nitrogen corresponded to 85 per cent, in contrast with only 1 per cent as ammonia. (See Florkin [31].)

4. Difficulties in the determination of the pattern of end products of amino acid metabolism

While referring to the introduction, at the eve of the nineteenth century, in the field of the metabolism of amino acids, of the concepts of detoxicating excretion syntheses expressed by the concepts of ureotelism and uricotelism as opposed to their lack in ammoniotelism, we may consider a number of obstacles to a correct recognition of such systematic characters that appeared later. The cases referred to above are clear-cut examples of the predominance of an excretory synthetic product determined in the excretory fluid produced by the nephridia, a procedure that gives the impression of the hypotheses standing on firm ground. It

* Prout was so surprised by this observation that he suggested that the snakes may have become sick in captivity.

References p. 207

would be misleading, nevertheless, to take, for instance, the distribution of nitrogenous compounds present in the urine of the carp (*Cyprinus carpio*) as indicating the relative proportions of the different nitrogenous excretion products, since some are known to be eliminated by the gills. In a number of cases, as expressed in tables printed in textbooks of comparative physiology, and in collections of biological data, the analysis of excretion products has been accomplished on the water in which the animals are kept and into which they eliminate not only the products of their nephridia, but also their feces.

A stimulation to necessary caution is found in the history of the interpretation of the nitrogenous excreta of the Roman snail (*Helix pomatia*). According to Hesse [32], *Helix pomatia* excretes 3.85 mg of nitrogen per kg. When this observation was made, it had long been known that the whitish content of the nephridia was for the greater part composed of uric acid, as L. Jacobson (cited by Wolf [33]) had established as early as 1820, and many authors confirmed. Marchal [34], having isolated uric acid from the nephridia, showed that each nephridium contained more than 7 mg of uric acid. In fact, at the end of hibernation, as was shown by Baldwin and Needham [35], a snail's nephridium contains an average of 32 mg of uric acid, i.e. about three-quarters of the dry weight of the organ. In the tables currently compiled in treatises of comparative physiology and in collections of biological data before World War II, the data on *Helix pomatia* were those of Delaunay [36] (1927), publicized in a review [37] he published in 1931, which contained much valuable material. Delaunay, analyzing a water extract of the nephridia of *Helix pomatia*, was led to conclude that the Roman snail had two kinds of excreta: solid, composed mainly of purines, and liquid, composed mainly of urea and ammonia. The question has been carefully revised by Jezewska, Gorowski and Heller [38], in the laboratory of Heller in Warsaw. These authors showed that 90 per cent of the total nitrogen found in the nephridia and in the nephridial excreta, separated from the feces, consisted of uric acid, xanthine and guanine. Urea, ammonia and allantoin were not found. The same authors showed that the evaluation of nitrogen compounds found in the water in which Delaunay had kept the snails partially immersed, amounted only to about 1 per cent of the total nitrogen of the excreta. The purinotelic character of the nitrogen metabolism of the Roman snail is obvious and the proportion in the excreta of end-products other than uric acid, guanine and xanthine, amounts to mere traces.

This example illustrates the difficulties inherent in a correct diagnosis of the nature of the end-products of amino acid metabolism. Fortunately the contrast between ureotelic mammals and uricotelic birds, which became the basis of the first biochemical studies on ureogenesis and uricogenesis, was a sound concept that was confirmed in subsequent studies.

5. The "law of arginase"

Introducing a number of ureides (guanidine, creatinine, arginine, allantoin, purine or pyrimidine derivatives, galegine) in mammalian organisms, Thompson [39] made the observation that among those compounds, arginine was the only one leading to an additional excretion of urea. This was in harmony with the recent discovery of arginase in intestinal mucosa and in other organs by Kossel and Dakin [40] who found that it catalysed the transformation of arginine into ornithine and urea. The background of this discovery consisted in observations made by Ch. Richet, between 1894 and 1897. (Literature in Kossel and Dakin [40].) Richet had observed the formation of urea as the product of digestion of liver tissue in water. The distribution of arginase in organisms has been the subject of a vast literature which has been recorded by Greenberg [41].

In 1913 and the beginning of 1914, Clementi was working in the laboratory of Kossel in Heidelberg. He was instructed by his chief to prepare a survey of the occurrence of arginase in the liver of different vertebrates. Clementi concluded that the presence of arginase was linked to the existence of an excretion of the nitrogen of amino acids in the form of urea (ureotelism), as it is in fishes, amphibians and mammals, whereas uricotelic vertebrates, such as reptiles and birds had no liver arginase. As World War I broke out in 1914 between Germany and Italy, Clementi [42,43] published the results in his own country and formulated what became known as the "law of arginase". Later, in 1915, Edlbacher, another collaborator of Kossel, published a paper [44] in which he agreed fairly well with Clementi's views. Paradoxically, in the

References p. 207

bellicose climate of the time, this confirmation was considered by Clementi as an act of war: as he later declared [45]:

"Nella seconda metà del 1915, in seguito alla dichiarazione di guerra dell'-Italia a l'Austria, Edlbacher abbe la singolare pretesa di volere controllare con poche analisi i resultati delle mie sistematiche ricerche e confermo le mie conclusioni circa l'assenza dell'arginasi nel fegato degli uccelli e dei rettile e la sua presenza nel fegato dei pesci, degli anfibi e dei mammiferi."

Pursuing his enquiry, Edlbacher, using a more refined methodology, detected, in collaboration with Bonem [46], the presence of arginase in the liver of male fowl. Edlbacher and Röthler [47] confirmed this conclusion for the male and female fowl. These findings of arginase in bird livers released the wrath of Clementi, who was, and remained, an irrepressible enthusiast of his own eponymous fame for discovering "Clementi's law". When the present author [48] mentioned the observations of Edlbacher in a book on biochemical evolution, Clementi [49] voiced a protest:

"Debbo qui pubblicamente deplorare l' inqualificabile atto di scorrettezza, che confina con la disonesta scientifica, di Florkin. . ."

But, if it is true that the ureotelic animals are provided with arginase the significance of which became clear after the formulation of the "ornithine cycle" (the positive aspect of the arginase law), it is not true that uricotelic animals lack arginase, the negative aspect of the law. The presence of arginase has been demonstrated in the liver of birds, sometimes at high levels (literature in Brown [50]), and in snakes and lizards [51].

The relation between the nature of the end-products of the catabolism of amino acids and the nature of the enzymological tools of the different animal groups will be understood later, as we shall see. Nevertheless, the notion of a high arginase activity in the liver of vertebrates (mammals, amphibians and certain reptiles) which can synthesize urea remained for a time current as a new form of Clementi's law, and was a positive factor in research on ureogenesis by suggesting a connection that was taken into account by Krebs and Henseleit in their interpretation of the "ornithine effect". (Chapter 48.)

REFERENCES

1 W.A. Smeaton, Fourcroy, Chemist and Revolutionary, 1755—1809, Cambridge, 1962.
2 N.G. Coley, Ambix, 18 (1971) 69.
3 C.W. Scheele, Opuscula II, 1776, p. 73 (translation by T. Beddoes: The Chemical Essays of C.W. Scheele, London, 1786, pp. 201—216) (quoted after Coley).
4 T. Bergman, in Opuscula IV, 1776, p. 2321.
5 L.B. Guyton de Morveau, Encyclopédie Méthodique. Chimie, Pharmacie et Métallurgie, I, part i, Paris, 1786, pp. 407, 508.
6 A.F. Fourcroy, Ann. Chim. Phys., 16 (1793) 96.
7 W.H. Wollaston, Phil. Trans. Roy. Soc., 87 (1797) 386.
8 R.A. Peters, in J. Needham (Ed.), The Chemistry of Life, Cambridge, 1970, p. 192.
9 G. Pearson, Phil. Trans. Roy. Soc., 88 (1798) 15.
10 A.F. Fourcroy, Ann. Chim. Phys., 27 (1798) 225.
11 A.F. Fourcroy and N. Vauquelin, J. Soc. Pharm., 1798, 263; Ann. Chim. Phys., 30 (1799) 54; 32 (1799) 216.
12 D.L. Drabkin, Thudichum. Chemist of the Brain, Philadelphia, 1958.
13 F.M. Rouelle, Journal de Médecine, nov. 1773 (quoted after Werner [14]).
14 E.A. Werner, The Chemistry of Urea, London, 1923.
15 F. Kurzer and P.M. Sanderson, J. Chem. Ed., 33 (1956) 452.
16 A.F. Fourcroy and N. Vauquelin, Mém. Inst., 2 (1799) 431.
17 M. Crosland, The Society of Arcueil, A View of French Science at the Time of Napoleon I, London, 1967.
18 J. von Liebig, Pogg. Ann., 17 (1829) 389.
19 J. Rollo, An Account of Two Cases of the Diabetes Mellitus, London, 1797.
20 A.F. Fourcroy and N. Vauquelin, Ann. Chim. Phys., 32 (1799) 80.
21 A.F. Fourcroy and N. Vauquelin, Ann. Mus., 11 (1808) 226.
22 J.L. Proust, Ann. Chim. Phys., 14 (1820) 257. (The author's name reads M. Proust, meaning Monsieur Proust.)
23 V. Regnault, Ann. Chim. Phys., 69 (1838) 180.
24 A.F. Fourcroy and N. Vauquelin, Ann. Mus., 4 (1804) 329.
25 N. Vauquelin, Ann. Mus., 18 (1811) 82.
26 A.F. Fourcroy and N. Vauquelin, Ann. Mus., 17 (1811) 310.
27 N. Vauquelin, Ann. Chim. Phys., 2nd ser., 21 (1822) 440.
28 W. Prout, cited by Werner.
29 M.R. Mattice, Chemical Procedures for Clinical Laboratories, Philadelphia, 1936.
30 S. Toda and K. Taguchi, Z. Physiol. Chem., 87 (1913) 371.
31 M. Florkin, L'évolution du Métabolisme des Substances Azotées chez les Animaux (Actualités Biochimiques, No. 3), Paris, 1945.
32 P. Hesse, Z. Allg. Physiol., 10 (1910) 273.

33 G. Wolf, Z. Vergl. Physiol., 19 (1933) 1.
34 P. Marchal, Mém. Soc. Zool. France, 3 (1889) 31.
35 E. Baldwin and J, Needham, Biochem. J., 28 (1934) 1372.
36 H. Delaunay, Recherches Biochimiques sur l'Excrétion Azotée des Invertébrés, Thèse Sci. Nat., Paris, 1927.
37 H. Delaunay, Biol. Rev., 6 (1931) 265.
38 M.M. Jezewska, B. Gorowski and J. Heller, Acta Biochim. Polon., 10 (1963) 55.
39 W.H. Thompson, J. Physiol., 32 (1905) 137; 33 (1905) 106.
40 A. Kossel and H.D. Dakin, Z. Physiol. Chem., 41 (1904) 321; 42 (1904) 181.
41 D.M. Greenberg, in J.B. Summer and K. Myrbäck (Ed.), The Enzymes, vol. I, part 2, New York 1951, p. 893.
42 A. Clementi, Rend. R. Accad. Lincei, Classe Sci. Fis. Nat., 23 (1914) 517.
43 A. Clementi, Arch. Fisiol., 13 (1915) 189.
44 S. Edlbacher, Z. Physiol. Chem., 95 (1915) 81.
45 A. Clementi, Boll. Soc. Ital. Biol. Sper., 22 (1946).
46 S. Edlbacher and P. Bonem, Z. Physiol. Chem., 145 (1925) 69.
47 S. Edlbacher and H. Röthler, Z. Physiol. Chem., 148, 273.
48 M. Florkin, L'Evolution Biochimique, Paris, 1944 (translation by S. Morgulis, Biochemical Evolution, New York, 1949).
49 A. Clementi, Boll. Soc. Ital. Biol. Sper., 26 (1950) 391.
50 G.W. Brown Jr., Arch. Biochem. Biophys., 114 (1966) 184.
51 J. Mora, J. Martuscelli, J. Ortiz-Pineda and G. Soberon, Biochem. J., 96 (1965) 28.

Chapter 48

Ureogenesis

1. The precursors of urea in the mammalian body

That urea pre-existed in the nitrogenous constituents of the body (later called proteins) remained a current view during the first half of the nineteenth century. This theory, according to which no synthesis of urea occurred in organisms but only a liberation of a preformed structure, was one of the reasons for the lack of consideration, in the perspectives concerning "vital force", of the significance of Wöhler's synthesis. (See Chapter 12, p. 253.) Urea was not believed to be synthesized, and was considered as a mere catabolic product.

After the concept of albuminoid compounds, or proteins, was formulated (Chapter 5), Voit [1] introduced the notion of the nitrogen equilibrium establishing a direct relation between the ingested "protein" and the excreted urea. But nobody was able to derive urea from proteins by submitting the latter to oxidation, considered at the time as the universal mechanism of metabolic changes. It is true that amino acids could be separated from proteins, but from the point of view of organic chemistry, no relation could be detected between the skeleton of urea, N—C—N, and the amino acids known (glycine, leucine, aspartic acid, glutamic acid, tyrosine). A young Pole, M. Nencki, born in 1847 in Kielce, at that time in Russia, was a medical student in Berlin and, in parallel with the medical curriculum, worked for two years in Baeyer's laboratory. (On Nencki, see Bickel [2] and Niemierko [3].) Nencki published in 1869 with another student, Schultzen, a paper [4] dealing with the new hypothesis of a synthetic derivation of urea from amino acids.

The background of this work is to be found, as noted by Bickel,

Plate 161. Marceli Nencki.

in the then recent discovery of the existence of amino acids in blood as well as the nitrogenous equilibrium recognized by Voit [1] (1866). Schultzen and Nencki [4] fed a dog on a minimal protein diet until the constant daily excretion of urea was lowered to 4 g. Adding glycine to the diet, they observed a corresponding increase of nitrogen excretion in the form of urea, and no excretion of unchanged glycine. They observed the same result with leucine. A control was performed with acetamide, chemically related to glycine although not an amino acid.

$NH_2—CO—CH_3$	$NH_2—CH_2—COOH$
acetamide	glycine

Acetamide was recovered as such in the urine, whose urea content remained unchanged. These experiments gave support to the concept of amino acids as precursors of urea in the mammalian body.

Nencki got his M.D. degree in 1870 with a dissertation on the oxidation of aromatic compounds in the body. Shortly afterwards he was called as assistant to the University of Bern where, a few years later, he became professor and head of the Department of Biochemistry. A number of different topics became the subjects of studies in Nencki's laboratory in Bern. One of them followed up his work on the precursors of urea. Nencki, recognizing that the glycine skeleton, N—C—C, is not related to the urea skeleton, N—C—N, was the first to consider urea not as directly derived from the catabolism of amino acids, but as a biosynthetic product. He found a model for such a synthesis in the formation of hippuric acid and in general in the conjugation of aromatic acids. In the conjugation of glycine with aromatic acids

$$Ar—C \rightarrow Ar—C—N—C—C$$

he recognized a model for a possible synthetic reaction of glycine with carbamic acid

$$N—C \rightarrow N—C—N—C—C$$

These studies led Nencki to consider the possibility of the accomplishment in organisms, not only of hydrolysis (splitting, with introduction of the elements of water) but also of synthesis by removal of the elements of water. He considered that an example of such synthesis was found in the formation of glycogen from glucose. He was led to reconsider the example of hippuric acid formation from benzoic acid and to confirm its accomplishment

References p. 226

by conjugation in the body, as it was known that the capacity of the dog was limited with respect to hippuric acid formation. Nencki selected dogs who did not excrete any hippuric acid. When, to such a dog, he gave 3 g of benzoic acid added to the diet, he observed that half of it appeared in the urine as hippuric acid and the other half as benzoic acid. Now, adding to the benzoic acid fed to the dog the corresponding amount of glycine, he identified all the benzoic acid in the form of hippuric acid. From these results, Nencki confirmed the recognition of the formation of hippuric acid as a conjugation of benzoic acid with glycine, limited by the amount of glycine available. He therefore recognized glycine as able to participate in synthetic reactions with removal of water.

2. Identification of urea in blood. Demise of the kidney as the site of urea formation

As urea had been isolated from urine, it was natural to believe it was formed in the kidney. The rupture with this concept took place when Prévost and Dumas [5] recognized the presence of urea in the blood. They removed the kidneys of a dog and observed an accumulation of urea in the blood. Prévost and Dumas quote from a book of Richerand [6] in which this author, who had removed the kidneys of dogs, observed that when the animals died, their gall bladder was always replete with fluid, a finding he considered as showing that when urea was not able to leave the body by way of the kidneys, it could leave it by another excretory path, a view shared by Prévost and Dumas. From their classical experiments, in which they combined vivisection methods and correct chemical analysis, Prévost and Dumas concluded that urea was not synthesized in the kidneys.

The discovery of Prévost and Dumas was recognized as fundamental by Magendie [7], who considered it "of the highest importance". It was confirmed by Segalas [8] and by Mitscherlich, Gmelin and Tiedemann [9]. The latter authors, after removal of the kidneys, found urea in the blood, but not in other body fluids. After the publication of the paper of Prévost and Dumas, Wöhler [10] commented on it and stated that it would appear improbable that an organ would take the place of another organ in the role of producing urea but that it was logical to believe that another organ could take the place of the kidney in accomplishing the function of removing urea from the blood. Wöhler considered that, for instance the appearance, in patients with impaired renal excretion, of a urinous smell of the breath, was in favour of the conclusion of Prévost and Dumas, that the kidney did not form urea.

Bernard and Barreswill [11] recognized that the substitute for urea secretion in cases of kidney removal and trapping of urea in blood was an excretion of ammonia by the digestive juices. This conclusion was obtained from the elegant removal of the kidneys of a dog with a gastric fistula. Bernard and Barreswill, starting from the classical research of Prévost and Dumas, recognized the equivalence of ammonia excretion considered as a product of fermentation in the stomach. As Holmes [12] remarks (p. 250, *Note* 50), the equivalence of urea and ammonia had been emphasized by Dumas in his *Essai de Statique Chimique* [13] (p. 37)

Bernard, like many biologists at the time (see section 4), had been interested in the process of secretion in general and he had been looking for substantiation of the concept according to which each secretion results from an alteration of a constituent of the blood, resulting in the formation of a specific secretory product. To quote Holmes [12] (p. 334):

"Now, however, he and Barreswill regarded the transformation as accidental. They believed that the urea itself is secreted into the stomach, where it is swept into the decomposing influences that nutrients and many other substances introduced there undergo. They supported this interpretation by showing that urea placed in the stomach of a living animal, or in contact with an intestinal membrane, was also converted to ammonia".

3. Influence of the work of Prévost and Dumas on the theories of secretion in general

In the frame of classical mechanical physiology, a gland was considered as extracting from the blood a preformed product, such as bile or urine. In vol. X of his *Système des Connaissances Chimiques*, published in 1801, Fourcroy [14] repudiated such views and ascertained that secretion must be a chemical phenomenon. This point was also made by Berzelius who proposed the distinction between secretion and excretion. Berzelius [15] called secretions the products which were, in a later phase, of use to the organism, while he called excretions those leaving the body, such as urine, sweat or milk.

It was the result of the work of Prévost and Dumas [5] which brought clarity to this confused domain. After they had observed, in cats and dogs, that, after removal of both kidneys, the urea concentration increased in the blood, it was clearly recognized that the kidney was not the organ of urea formation, but the elimination site of the urea formed at another site. To these experiments of Prévost and Dumas, Magendie [7] refers, as proving that the kidney does not form urea

"as one generally used to believe, but that they simply separate it from the blood, where it forms". (Translation by Holmes [12], p. 335.)

That the urea content of blood and tissues is independent of the kidney was confirmed by a number of authors. (Literature in Bollman, Mann and Magath [16].)

In his *Handbuch* (vol. I, p. 585) Müller [17] asks

"Whether there is another specific organ, or whether it forms in various tissues, whether it results only from the decomposition of the substance of the animal or also directly from unused nutrients". (Translation by Holmes [12], p. 335.)

The opinion that urea forms in the blood, as suggested by Magendie, was not shared by Bernard. Holmes, after perusing Bernard's laboratory notebook, reached the conclusion that Bernard apparently had in mind the hypothesis according to which the general lymphatic circulation carried urea from the site of its formation to the blood. In this line, Holmes has retraced the trail of research followed by Bernard and the reasons for his failing to reach a conclusion. Bernard's investigations seem

"to have embodied the goal of substantiating physiologically the theory that urea is the decomposition product of the organized nitrogenous constituents of the body tissues" [12] (p. 349).

In this issue, Bernard was endeavoring to answer one of the queries of Müller.

This idea that the formation of urea gives a measure of the decomposition of the nitrogenous constituents of organized tissues had been formulated by Liebig [18] in his *Animal Chemistry*. Marchand [19], in 1838, published results pointing to such origin for at least a part of the urea excreted. Marchand observed that, even with no nitrogen in the diet of a dog, urea excretion continued.

4. Urea formation in the liver

A number of experimenters attempted to determine the site of formation of urea by comparing, for example, the urea content of peripheral blood with that of the hepatic vein. (Literature in Bollman et al. [16].) As in many aspects of physiological chemistry studied on whole organisms, the chemical methods used at the time lacked the specificity and the accuracy that would have been necessary to reach a firm conclusion.

Another approach was based on a comparison of the urea content of the tissues. That the liver contains considerable amounts of urea was discovered by Heynsius and confirmed by Stokvis. (Literature in Bollman et al. [16].) Stokvis suggested that urea, in the mammalian liver, was formed by oxidation of uric acid. This was in agreement with the old ideas according to which catabolism involved oxidation. Meissner [20], in 1860, had found urea in the liver of dogs, and uric acid in the liver of hens, both in concentrations higher than in blood. In 1890, Nencki was called to St. Petersburg to participate in the organization of an Imperial Institute for Medical Research. He was appointed head of the Department of Chemistry and Biochemistry, while Pavlov, two years younger than Nencki, was the head of the Department of Physiology.

Nencki returned, partly with Pavlov, to the study of the site of urea formation in the body. Later on, when Nencki was attracted towards other studies (for instance his well-known work on haemoglobin), the investigations on urea were prolonged in his laboratory by his coworkers Salaskin, Zaleski and others. Nencki's idea was experimentally to modify the conditions of the liver in relation to the organism, and it is in this respect that the help of Pavlov's team allowed for several methods of interfering more or less with the blood supply of the liver. Hahn, Massen, Nencki and Pavlov [21] cut off the portal blood supply to the liver by performing an Eck's fistula and, in addition, they ligated the hepatic artery. In some experiments, they partially extirpated the liver. The inconvenient aspect of these experiments was the short survival of operated dogs. The authors observed that the portal blood may contain three to five times more ammonia than the blood of the general circulation. After the establishment of an Eck's fistula, they did not observe any significant increase of ammonia in the general circulation. When, in addition to an Eck's fistula, they ligated the hepatic artery and removed an important part of the liver, the ammonia increased in the systemic blood system, giving rise to ammonia poisoning The increase of ammonia in the blood was repeatedly confirmed by different authors in spite of their disagreements on the part of ammonia in the production of the symptoms. (Literature in Mattews and Miller [22].) These experiments proved the implication of the liver in the biosynthesis of urea from metabolic ammonia.

While the mechanism of the synthesis remained in the dark, it was confirmed that the amino acids were the precursors, in different tissues, of ammonia or carbamic acid which took part in a ureogenesis that the authors considered as taking place in the liver, though they did not exclude its accomplishment in other organs. On the other hand, the collaboration of Nencki and Pavlov in St. Petersburg opened the way for the developments of experimental surgery described below.

Another approach was accomplished by the method of the perfused liver. Von Schröder [23], in Schmiedeberg's laboratory, observed that, in perfused surviving liver, urea is produced from ammonium carbonate, and this result was confirmed by other authors. In such experiments, the amount of urea nitrogen produced approximately corresponded to the ammonia nitrogen introduced (Löffler, [24]). But it was Salaskin [25], in Nencki's laboratory, who was the first to demonstrate that a surviving mammalian liver perfused with blood containing added glycine or aspartic acid was the seat of urea production. This conclusion was denied by some, but confirmed by those who managed to maintain good tissue arterialization. (Literature in Fearon [26].) If such experiments confirmed the ureogenesis in liver, they brought no information about the possibility of its accomplishment in other organs as suggested by the experiments of Nencki and Pavlov. Perfusion experiments showed that if other organs (kidney, pancreas, intestinal mucosa) were perfused with blood containing added amino acids, the nitrogen of these compounds was liberated in the form of ammonia (Lang [27], Bostock [28]). Though subject to the reservation with which experiments on more or less oxygenated but surviving tissues were interpreted, the results pointed to the notion of the specialization of the mammalian liver in the biosynthesis of urea from the ammonia resulting from deamination of amino acids. (See Chapter 35.)

5. Experiments with amino acids on whole animals

The first experiments on the fate of amino acids derived from the hydrolysis of proteins when injected into the body were accomplished by Nencki and his friend Schultzen [29], who observed that practically all the nitrogen was eliminated as urea and to a lesser extent as ammonia, whereas the carbon of the amino acid is eventually oxidized to carbon dioxide (confirmed by Salkowski [30]).

In 1903, Stolte [31] published results of experiments and concluded that after amino acids had been injected intravenously into a rabbit, glycine and leucine yielded urea almost completely, whereas other amino acids, such as phenylalanine or tyrosine appeared as such in urine, and no supplementary excretion of urea was observed. These results were not in accord with those of other authors. It is only when large doses of amino acids were injected that small amounts were recovered in urine. For instance, Salkowski [32], having given 25 g of glycine to a dog, recovered a small part in urine. But, when using smaller doses, other experimenters (Abderhalden [33—35], Loewy and Neuberg [36]) found that, in normal mammalian organisms, the amino acids were completely metabolized to urea, a concept that has been current since.

The participation of the liver in this function was approached by experiments of non-destructive nature performed on whole normal mammals. Van Slyke and Meyer [3], having previously observed that amino acids injected into the circulation of dogs are absorbed by the tissues, showed that the absorbed amino acids rapidly disappear from the liver. In this organ, the amino nitrogen content may be doubled by injecting amino acids into the circulation and yet return to the normal level after 2 or 3 h. During the period in which the liver gets rid of the absorbed amino acids, the concentration, in muscle for instance, does not fall. The disappearance of amino acids from the liver is accompanied by an increased concentration of urea in the blood.

References p. 226

6. More efficient surgery

From the observations described above, we may conclude that urea is formed in the mammalian liver. But the same observations, except some results obtained on surviving perfused liver, do not exclude a formation of urea from ammonia in other parts of the mammalian organism. This point was cleared up by Bollman, Mann and Magath [16], at the Mayo Clinic, Rochester, Minnesota, after perfecting the surgical procedures that had been relied upon by Nencki and Pavlov to modify the relationship of liver and organism. Mann [38] published (1921) an experimental method in which total removal of the liver from dogs allowed the animals to remain alive and normal to all outward appearances, for 24 h. In this method, the liver was removed rapidly with little loss of blood and the other tissues did not suffer.

To quote from Bollman et al. [16]:

"The very striking decrease in the urea content of the blood and tissues, with the equally marked decrease in the urea excretion, proves conclusively that there is a marked decrease in the amount of urea formed in the body following removal of the liver. The progressive decrease in urea to such a minimal amount in both blood and urine is so marked as to indicate strongly that urea formation ceases immediately following hepatectomy".

From this "strong indication", Bollman, Mann and Magath [16] concluded that

"the production of urea in the body of the dog is entirely dependent on the presence of the liver, since urea formation ceases completely as soon as the liver is removed".

The recognition of the mammalian liver as the exclusive site of ureogenesis had already been substantiated by perfusion experiments on isolated organs, and it won further recognition through a host of later enzymological results.

7. Chemical schemes for urea formation

After it was recognized that the first step in the formation of urea from amino acids in mammals was the liberation of ammonia and of carbon dioxide, the theory was proposed that a carbamate was converted into urea (von Schroeder [23]).

Ammonium carbamate

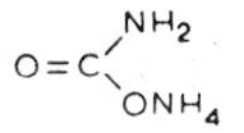

$$O{=}C\begin{matrix} \diagup NH_2 \\ \diagdown ONH_4 \end{matrix}$$

was considered to be transformed into urea and water, a reversal of the well-known hydrolysis of urea by boiling or by heating in acids or in alkaline media in vitro

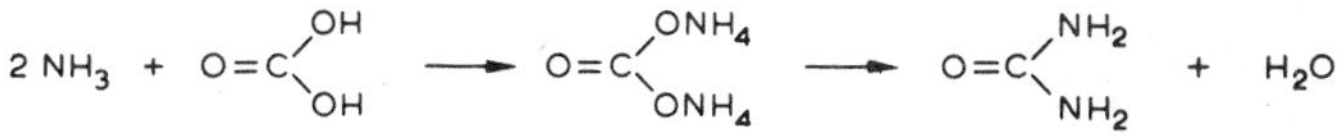

$$2\,NH_3 + O{=}C\begin{matrix} \diagup OH \\ \diagdown OH \end{matrix} \longrightarrow O{=}C\begin{matrix} \diagup ONH_4 \\ \diagdown ONH_4 \end{matrix} \longrightarrow O{=}C\begin{matrix} \diagup NH_2 \\ \diagdown NH_2 \end{matrix} + H_2O$$

Other chemical theories have been proposed. (Literature in Werner [39] and in Neubauer [40].)

It has, for instance, been suggested by Drechsel [41] that ammonium carbamate was an intermediary step between ammonium carbonate and urea.

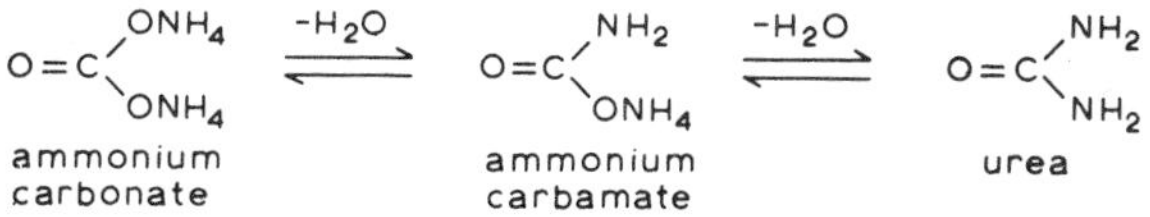

$$\underset{\text{ammonium carbonate}}{O{=}C\begin{matrix} \diagup ONH_4 \\ \diagdown ONH_4 \end{matrix}} \underset{}{\overset{-H_2O}{\rightleftharpoons}} \underset{\text{ammonium carbamate}}{O{=}C\begin{matrix} \diagup NH_2 \\ \diagdown ONH_4 \end{matrix}} \overset{-H_2O}{\rightleftharpoons} \underset{\text{urea}}{O{=}C\begin{matrix} \diagup NH_2 \\ \diagdown NH_2 \end{matrix}}$$

This theory was accepted by a number of authors (see Neubauer [40]), but it was contradicted by Löffler [24] on the basis of the occurrence of urea formation in the isolated liver perfused with acid solutions.

A theory of urea formation by an oxidative process was formulated by Hofmeister [42]. It was based on a test-tube reaction: the oxidation by permanganate, of proteins, amino acids, or even N-free compounds, in the presence of ammonia, resulting in the formation of urea. Several authors have interpreted this reaction as taking place without an oxidative formation of urea, and have expressed the view that it was not in contradiction with the ammonium carbamate theory. Another theory proposed by Hoppe-Seyler [43] and supported in the book of Werner [39], was derived from another test-tube reaction, the production of urea after the addition of ammonia to a cold solution of cyanic acid, a reaction similar to the classical synthesis of Wöhler. This theory was widely accepted before the discovery of the ornithine cycle.

Later, Fosse [44] combined the Hoppe-Seyler cyanic acid theory with the oxidation theory of Hofmeister, and on the basis of test-tube reactions he proposed that urea was formed by the

following reactions:

(1) $HCHO + NH_3 + O \rightarrow HCN + 2\ H_2O$

formaldehyde (derived from carbohydrates) — cyanhydric acid

(2) $HCN + O \rightarrow CONH$

cyanic acid

(3) $CONH + NH_3 \rightarrow CON{-}NH_4 \rightarrow CO(NH_2)_2$

cyanic acid — ammonium cyanate — urea

But the ammonium carbamate theory prevailed since it was considered as having received strong support from the classical experiments of Schröder [23] on the isolated perfused liver.

8. The "ornithine effect"

As has been generally observed, the method of experiments on whole animals and even on isolated organs, was rarely able to lead to unequivocal final conclusions concerning metabolic pathways. Krebs had worked between 1926 and 1930 in the laboratory of Warburg where he was impressed by the potentialities of the tissue-slice technique developed by Warburg [45] in his studies on the catabolic processes of glycolysis and respiration. When he joined, in 1930, the Department of Internal Medicine of Freiburg University, Krebs [46,47] conceived the idea of relying on the tissue-slice technique for a study of biosynthetic processes, as tissue slices were able to use for such processes the energy they were known to derive from respiration and glycolysis. To test this idea, Krebs chose the biosynthesis of urea by the mammalian liver. Krebs was led to the selection of this particular biosynthesis by the high rate of urea synthesis in whole organisms. (In man, on an ordinary diet, 30 g of urea are produced in the course of 24 h; and with a high-protein diet the rate may be trebled.) Krebs, before testing liver slices, adapted rapid methods of urea determination to the conditions of manometry, and designed a new saline providing as completely as possible the conditions obtaining in the natural

References p. 226

environment of cells. Krebs accomplished preliminary tests with tissue slices and observed that they carried out urea synthesis at the expected rate which increased after the addition of ammonium salts or certain amino acids.

To quote Krebs [47]:

"After this hopeful start I decided to measure systematically the rate of urea synthesis in the presence of a variety of precursors hoping that the results may throw light on the chemical mechanism of urea synthesis. I had no preconceived idea and no hypothesis about this mechanism but I had in mind Warburg's work on cell respiration in which he had studied the rate of respiration in the presence of cyanide, carbon monoxide, narcotics and other substances and succeeded in drawing far-reaching conclusions about the proper-

Plate 163. Kurt Henseleit and Sir H.A. Krebs.

ties of the enzyme of respiration. In this work I was joined by a medical student, Kurt Henseleit, who had asked my chief, Prof. Thannhauser, to suggest a subject for his M.D. thesis. Thannhauser referred him to me. He proved a very skillful experimenter who rapidly acquired the necessary laboratory techniques" (p. 20).

In the tissue-slice method, a large number of experiments could be performed under well controlled conditions. Krebs and Henseleit [48] planned an outline of experimental researches involving a number of queries. In the conversion of amino nitrogen, is ammonia a necessary intermediate? How do the rates of urea formation from different amino acids compare? Can cyanate, considered as an intermediate in a currently accepted theory, be converted to urea? Krebs and Henseleit, in approaching such queries, measured the rate of urea formation by rat-liver slices in different conditions, for instance with various mixtures of ammonium ions and different amino acids. In the course of these experiments, they were surprised to observe an exceptionally high rate of urea biosynthesis when both ornithine and ammonium ions were present ("ornithine effect"). This striking enhancement of urea formation was the starting point of a highly productive enquiry.

9. Analysis of the "ornithine effect"

Being aware of the existence of arginase, Krebs and Henseleit [48] at first doubted the ornithine effect and suspected a contamination of ornithine by arginine. This possibility was discarded. They thought of the possibility of an articulation of the ornithine effect with the action of arginase, but without first visualizing the nature of such an articulation. Testing low concentrations of ornithine, they observed the same stimulating effect as with high concentrations. Moreover they observed that the ornithine concentration did not decrease during the effect, and that the total urea produced was accounted for by the lowering of the ammonium salts. Obviously, ornithine acts like a catalyst, and the authors were led to recognize that arginine had to be considered as an intermediate:

$$\text{ornithine} + CO_2 + 2NH_4^+ \rightarrow \text{arginine} .$$

Arginine was also found to stimulate the synthesis. So Krebs began

References p. 226

to look for intermediates between ornithine and arginine. From the consideration of its chemical structure, Krebs considered as a possible intermediate a compound, citrulline, which had just been isolated by two different researchers: by Wada [49] from water melon (*Citrullus*) and by Ackermann [50] in the bacterial degradation of arginine. To quote Krebs [47]:

"I wrote to both and obtained a few milligrams from each, sufficient to do the decisive tests. These entirely fulfilled expectations: they demonstrated the rapid formation of urea in the presence of citrulline and ammonium salts in accordance with this scheme:

$$\underset{\text{citrulline}}{COOH{\cdot}CH(NH_2){\cdot}CH_2{\cdot}CH_2{\cdot}CH_2{\cdot}NH{\cdot}CO{\cdot}NH_2}$$

$$\downarrow + NH_3$$

$$\underset{\text{arginine}}{COOH{\cdot}CH(NH_2){\cdot}CH_2{\cdot}CH_2{\cdot}CH_2{\cdot}NH{\cdot}C}\begin{matrix}{\diagup\!\!\diagup} NH\\ \diagdown NH_2\end{matrix} \quad + \quad H_2O$$

$$\downarrow + H_2O \quad \text{(arginase)}$$

$$\underset{\text{ornithine}}{COOH{\cdot}CH(NH_2){\cdot}CH_2{\cdot}CH_2{\cdot}CH_2NH} \quad + \quad \underset{\text{urea}}{NH_2{\cdot}CO{\cdot}NH_2}$$

"On the basis of these findings, it became possible to formulate a cyclic process of urea formation from carbon dioxide and ammonia, with citrulline and arginine as intermediate stages as shown below:

ornithine → citrulline ($+ CO_2$, $+ NH_3$, $- H_2O$)

citrulline → arginine ($+ NH_3$, $- H_2O$)

arginine → ornithine + urea ($+ H_2O$)

10. Ornithine cycle and "ureotelism"

In the Amphibia, as well as in mammals, the "ornithine cycle" is operative, as was demonstrated with liver slices in vitro for *Rana esculenta*, *R. temporaria* and *Bufo vulgaris* by Manderscheid [51] and by Münzel [52]. The same authors also detected the operation of the ornithine cycle in the liver of turtles and tortoises. On the other hand, no evidence for the cycle was recognized by Manderscheid in slices of hen's liver. Needham, Brachet and Brown [53] did not detect it in chicken embryos. Neither was the operation of the ornithine cycle recognized in liver slices of a lizard (*Lacerta serpa*) by Münzel [52].

At the time, the interpretation of the presence or lack of the operation of the ornithine cycle, subserving ureotelism, was formulated, not in terms of presence or lack or arginase, but in terms of greater or smaller concentration in the liver, a concept which would be replaced after 1946, by more adequate notions when a new formulation of the ornithine cycle resulted, as we shall see, from the introduction of carbamoyl phosphate ($NH_2 \cdot CO \cdot O \cdot PO_3H_2$) and arginino-succinate as intermediates.

11. The ornithine cycle as a pattern of metabolic organization

In the midst of rather dull approaches to biosynthesis during the period preceding World War II, marked mainly by static collections of data obtained by the methods of classical analytical chemistry (pipettes and burettes) and rather crude enzymology, the discovery of the "ornithine cycle" came as a most impressive achievement, opening new vistas on the dynamics of the complexities of biosynthesis. As stated by Ratner [54], the new concept introduced by the discovery of the "ornithine cycle" was

> "the accomplishment of a metabolic synthesis through the agency of carrier compounds by the repeated turnover of a cycle".

Based on studies on tissue slices, the existence of the pathway in whole animals was confirmed through some of the first applications of isotopic methods, by the use of ^{15}N. (Foster, Schoenheimer and Rittenberg [55]; Clutton, Schoenheimer and Rittenberg [56]; Evans and Slotin [57]; Rittenberg and Waelsch [58].)

References p. 226

REFERENCES

1 C. von Voit, Physiologische Untersuchungen, Augsburg, 1857.
2 M.H. Bickel, Marceli Nencki 1847–1901, Bern, 1972.
3 W. Niemierko, in C.C. Gillispie (Ed.), Dict. Sci. Biogr., vol. 10, New York, 1974, p. 22.
4 O. Schultzen and M. Nencki, Z. Biol., 8 (1872) 124.
5 J.L. Prévost and J.B. Dumas, Ann. Chim. Phys., 23 (1823) 94.
6 A. Richerand, Nouveaux Eléments de Physiologie, vol. II, Paris (Quoted by Prévost and Dumas). [5]
7 F. Magendie, Précis Elémentaire de Physiologie, 3rd ed., Paris, 1833.
8 P.S. Ségalas d'Etchepare, J. Physiol., 2 (1822) 354.
9 E. Mitscherlich, L. Gmelin and F. Tiedemann, Z. Physiol., 5 (1833) 1.
10 F. Wöhler, Z. Physiol., 1 (1824) 311.
11 C. Bernard and C. Barreswill, Arch. Gén. Méd., 13 (1847) 449.
12 F.G. Holmes, Claude Bernard and Animal Chemistry, The Emergence of a Scientist, Cambridge, Mass., 1974.
13 J.B. Dumas, Essai de Statique Chimique des Etres Organisés, Paris, 1841.
14 A.F. Fourcroy, Système des Connaissances Chimiques, vol. X, Paris, 1801.
15 J.J. Berzelius, Med.-Chir. Trans., 3 (1812) 234.
16 J.L. Bollman, F.C. Mann and T.B. Magath, Am. J. Physiol., 69 (1924) 371.
17 J. Müller, Handbuch der Physiologie des Menschen für Vorlesungen, 3rd ed., vol. I, Coblenz, 1838.
18 J. Liebig, Animal Chemistry (translated by W. Gregory), Cambridge, 1842.
19 R.F. Marchand, J. Prakt. Chem., 14 (1838) 496.
20 G. Meissner, cited by Bollman et al.
21 M. Hahn, O. Massen, M. Nencki and I. Pavlov, Arch. Exp. Pathol. Pharmakol., 32 (1893) 121.
22 S.A. Matthews and E.M. Miller, J. Biol. Chem., 15 (1913) 87.
23 W. von Schröder, Arch. Exp. Pathol. Pharmakol., 15 (1882) 364; 19 (1885) 373.
24 W. Löffler, Biochem. Z., 85 (1918) 230.
25 S. Salaskin, Z. Physiol. Chem., 25 (1898) 128.
26 W.R. Fearon, Physiol. Rev., 6 (1926) 399.
27 S. Lang, Beitr. Chem. Physiol. Pathol., 5 (1904) 321.
28 G.R. Bostock, Biochem. J., 6 (1911) 48.
29 O. Schultzen and M. Nencki, Ref. 4 and Ber. Dtsch. Chem. Ges., 2 (1869) 566; Biol., 8 (1872) 124.
30 E. Salkowski, Ber. Dtsch. Chem. Ges., 7 (1875) 116.
31 K. Stolte, Beitr. Chem. Physiol. Pathol., 5 (1903) 15.
32 E. Salkowski, Z. Physiol. Chem., 4 (1880) 54, 100.
33 E. Abderhalden and J. Markwalder, Z. Physiol. Chem., 72 (1911) 63.
34 E. Abderhalden, A. Furns, E. Goebel and P. Strübel, Z. Physiol. Chem., 74 (1911) 481.
35 E. Abderhalden and P. Bergell, Z. Physiol. Chem., 39 (1903) 9.
36 A. Loewy and C. Neuberg. Z. Physiol. Chem., 43 (1904)

37 D.D. van Slyke and G.M. Meyer, J. Biol. Chem., 6 (1913–1914) 213.
38 F.C. Mann, Am. J. Physiol., 55 (1921) 285.
39 E.A. Werner, The Chemistry of Urea, London, 1923.
40 O. Neubauer, in A. Bethe, G.V. Bergmann, G. Embden and A. Ellinger (Eds.), Handbuch der Normalen und Pathologischen Physiologie, vol. 5, Berlin, 1928.
41 E. Drechsel, J. Prakt. Chem., 12 (1875) 417.
42 F. Hofmeister, Arch. Exp. Pathol. Pharmakol., 33 (1894) 198.
43 F. Hoppe-Seyler, Physiologische Chemie, Berlin, 1871.
44 R. Fosse, Ann. Inst. Pasteur, 34 (1920) 715.
45 O. Warburg, Biochem. Z., 142 (1923) 317.
46 H.A. Krebs, in J.B. Summer and K. Myrbäck (Eds.), The Enzymes, vol. 2, part 2, New York, 1952, p. 866.
47 H.A. Krebs, Biochem. Educ., 1 (1973) 19.
48 H.A. Krebs and K. Henseleit, Z. Physiol. Chem., 210 (1932) 33.
49 M. Wada, Biochem. Z., 224 (1930) 420.
50 D. Ackermann, Biochem. Z., 203 (1931) 66.
51 H. Manderscheid, Biochem. Z., 263 (133) 245.
52 P. Münzel, Zool. Jahrb. (Physiol.), 59 (1938) 113.
53 H. Needham, J. Brachet and R.K. Brown, J. Exp. Biol., 12 (1935) 321.
54 S. Ratner, Adv. Enzymol., 15 (1954) 319.
55 G.L. Foster, R. Schoenheimer and D. Rittenberg, J. Biol. Chem., 127 (1939) 319.
56 R.F. Clutton, R. Schoenheimer and D. Rittenberg, J. Biol. Chem., 132 (1940) 227.
57 E.A. Evans Jr. and L. Slotin, J. Biol. Chem., 136 (1940) 805.
58 D. Rittenberg and H. Waelsch, J. Biol. Chem., 136 (1940) 799.

Chapter 49

Uricogenesis and Uricolysis

1. Origin of the uric acid excreted by birds

Liebig, as noted in Chapter 7, recognized as end-products of animal tissue metabolism the nitrogenous constituents of urine (urea and uric acid) and the carbonaceous components of bile. In his theory, they resulted from slow combustion (oxidation) which divided the proteinaceous constituents into two parts, one containing mainly nitrogen and one containing mainly carbon. Urea he considered as a product of further oxidation of uric acid, and the predominance of the latter in the excreta of birds and reptiles was interpreted by him as the result of a deficient oxidation, in animals that "seldom drink".

Doubts were expressed about the deficient oxidation theory by Salkowski [1] in 1877. Cech [2], in Salkowski's laboratory, the chemical section of the Institute of Pathology in Berlin, showed that urea fed to hens is not recovered as urea in the excretions. During the same year, von Knieriem [3] showed that over 90 per cent of ingested nitrogen in the form of glycine, leucine or asparagine, was excreted by hens as uric acid. During the following year, von Schroeder [4] fed hens with amino acids and with ammonium salts and showed that they can be converted into uric acid.

The synthetic theory of the origin in protein metabolism of the uric acid excreted by birds was established in 1886 by the classical experiments of Minkowski [5], who showed that, after extirpation of the liver, uric acid no longer occurs in the excreta of geese. In these conditions the birds excreted ammonium lactate. Minkowski concluded that the uric acid excreted by birds is derived from amino acids and represents an excretion synthesis for the "detoxication" of ammonia.

References p. 263

Plate 164. Oskar Minkowski.

The uricotelism of birds would probably have aroused little attention from the quarters of animal chemistry and pathology if it had not been known at the time that one of the symptoms of a human disease, gout, was an increase in concentration of blood uric acid. At a period when the nucleic acids were still unknown, it was the interest in gout that made the scene conducive to studies in uricotelism, which would probably have been classified otherwise among natural curiosities.

2. Gout

We have referred above to the discovery, by Wollaston, of uric acid in a chalky deposit (tophus) in gout. On February 8, 1848, at the meeting of the Royal Medical and Chirurgical Society of London, Alfred Baring Garrod (later Sir Alfred) *, professor of Therapeutics at King's College Hospital, London, reported on the application of a micro-test for uric acid, which allowed him to determine approximately the amount of uric acid in blood serum. (On A.B. Garrod, see Peters [6], and Rutz [7].) Peters (p. 197) considered this method

"as a precursor of modern chromatographic procedures".

Garrod's observations appeared in the form of a book [8] in 1860. In this book, Garrod gave a detailed description of the method. Into a flat glass dish, after adding acetic acid, one or two fibres, about an inch in length, were introduced

"from a piece of unwashed huckaback or other linen fabric, which should be depressed by means of a small rod, as a probe or the point of pencil".

After the dish has been left for two or three days in a cool place, when there is much uric acid, it crystallizes on the fibre.

"assuming forms not unlike that presented by sugar-candy upon a string".

* A.B. Garrod was the father of Alfred Henry Garrod (1846—1879), and of Sir Archibald E. Garrod (1857—1936) the author of *Inborn Errors of Metabolism* (London, 1909).

Plate 165. Sir Alfred Baring Garrod.

3. Xanthic bases

One of the first manifestations of the interest of organic chemists in biological problems was found in the study of the chemistry of xanthic compounds. As stated in Chapter 12 (p. 261), this interest appeared in the work of Wöhler and Liebig (1838), a masterpiece of the systematic study of natural compounds, in which uric acid ($C_5N_4H_4O_3$) was chemically related to allantoin, alloxan, alloxantin and other compounds. Among Liebig's students, several organic chemists maintained the interest of the master [10]. These included J.J. von Scherer (1814—1869), who worked with Liebig in 1840—41 and who became professor in 1841.

Another chemist was A.F. Strecker (1822—1871) who was Liebig's assistant from 1846 to 1848 and became professor in Christiania in 1851. He later (1860) became professor at Tübingen and finally succeeded Scherer in Würzburg in 1870. One of his former assistants, Medicus, later (1900) professor in Würzburg, first proposed a structural formula for uric acid.

(a) Xanthine

Here again, urinary stones were at the origin of the isolation, in 1817, of the compound by Marcet [11], a Swiss doctor living in London, who called it xanthic oxide. Liebig and Wöhler obtained the empirical formula $C_5H_4N_4O_2$.

(b) Hypoxanthine

Hypoxanthine was first isolated by Scherer [12], at the time professor in Würzburg, from the spleen and blood of a leukaemic patient. Strecker [13], when he was professor in Christiania, isolated it in 1858 from muscles (human and lower species) and called it sarkin. Strecker [13] tried unsuccessfully to oxidize it to xanthine or to reduce it to uric acid. These transformations, confirming the kinship of the compound with uric acid and with xanthine, were accomplished indirectly by E. Fischer and directly by Sundwik [14] in 1911.

References p. 263

(c) Guanine

Guanine was first isolated in 1844 by Unger [15], a student of Magnus in Berlin, from guano, which had been introduced into Europe by Humboldt around 1840. Strecker [13], in the same publication in which he announced his finding of hypoxanthine in muscles, showed by oxidation of xanthine under the action of nitric acid, the kinship of guanine with the group of xanthic bases. Guanine was isolated in 1866 by Virchow [16] from muscle concretions of pigs suffering from guanine gout.

4. The murexide reaction

As stated above, Scheele had observed the formation of red or purple products when uric acid was treated with concentrated nitric acid. Coley [17] has retraced the history of this reaction in an interesting section of his paper on stones, and we shall closely follow his presentation and primary bibliography. A first approach to a study of the purple compounds was accomplished by Brugnatelli [18] who isolated an acid crystallizing in colourless transparent needles (erythric acid) whose salts (erythrates) were red or purple. Brugnatelli did nevertheless call attention to the complexity of the subject.

Experimenting in the same field, Prout [19], in 1818, obtained red crystals of a new compound, the ammonium salt of an acid which was called "purpuric acid" (at the suggestion of Wollaston). As he had been accused by Vauquelin of having plagiarized Brugnatelli, Prout [20], in 1819, suggested that his purpuric acid was different from Brugnatelli's erythric acid. This was based on his observation that by the action of nitric acid on purpuric acid, a compound similar to erythric acid was obtained.

In 1820, Prout described a pink sediment of urine, and suggested that the colour was due to ammonium purpurate, a view that was later challenged by Bird and Brett [21]. Prout's ammonium purpurate was rechristened by Liebig and Wöhler in their paper of 1838 as murexide, to which they attributed the empirical formula $C_{12}N_{10}H_{12}O_8$ (modern formula, $C_6N_6H_8O_6$, Partington [22] p. 333) and they gave the purpuric acid of Prout the name murexane. The empirical formula determined by them was C_6N_4-H_8O_5 (modern formula $C_4N_5H_5O_3$).

5. The alleged relation of xanthic bases with proteins

This was a current belief, until the eighties. It resulted from the isolation of hypoxanthine (sarkin) from muscle by Strecker (see above). This theory was reinforced by the alleged isolation of hypoxanthine from fibrin, claimed by G.Salomon in 1878 and confirmed by R.H. Chittenden during the following year (see Fruton [23], p. 189). As was shown by Kossel in 1881, this hypoxanthine was derived from leucocytes sticking to the fibrin clot.

6. Miescher and nuclein

Miescher (see Olby [24]) obtained his M.D. degree in 1868 in Basel where he was born and where his uncle Wilhelm His was professor of anatomy and physiology. Miescher went to study chemistry with Strecker in Tübingen. His motivation was a deep interest in the reduction of cytological facts, including the staining properties of cells, to chemical data. To isolate nuclei, he started, in Hoppe-Seyler's laboratory, a study on pus cells, secured by extracting surgical bandages. (This was the age of *pus bonum et laudabile*.) He removed the cytoplasm by a digestion with artificial gastric juice (extract of gastric mucosa), obtaining large amounts of isolated nuclei. From this material, Miescher obtained a solution by treatment with dilute sodium carbonate. Adding acetic acid to the solution, he obtained a precipitate containing phosphorus and giving protein colour tests. This phosphorylated acid protein, resisting the action of pepsin, was considered by Miescher as a chemical individual (or a mixture of closely related ones). This constituent of the cell nucleus he called "nuclein" [25].

Having qualified in Basel in 1868, Miescher, during the winter of 1869—1870, went to Ludwig's laboratory in Leipzig, to complete his training in physiology. In the beginning of 1871, he returned as instructor to Basel.

W. His, from 1870 to 1872, when he was succeeded in the chair of physiology by Miescher, had studied the embryology of salmon development. Situated on the Rhine in which, each year in November, innumerable salmons travelled up to spawn in fresh water, Basel was an ideal spot for such studies. It was known that sperm

Plate 166. Johann Friedrich Miescher.

heads are equivalent to cell nuclei, and the local fish dealers were accustomed to provide the department of His with roes replete with eggs or sperm.

Miescher noted the increase in weight of the gonads over a period of several weeks during which the salmon is fasting, and the corresponding intense synthesis of nuclear material that he considered as derived from muscle proteins. The spermatozoa are composed of head, tail and middle part, the two latter parts amounting to an insignificant mass, and the tails being soluble in acetic acid. From his studies, Miescher [26] concluded that the sperm heads were almost completely made of a single chemical substance. This he considered to be a salt of a "xanthin-alkaloid", a base rich in nitrogen which he called "protamine", and of an organic acid containing phosphorus and which he called nuclein. It may be noted that Miescher considered the substrate of the positive murexide reaction as included in its "protamine" and was inclined to consider his "nuclein" as a compound formed between proteins and phosphoric acid.

Miescher added to the confusion by calling the salt corresponding to the material of the head "impure nuclein" and the phosphorus material "pure nuclein".

Miescher, a student of Strecker, realized the importance of his observation of a positive murexide test at the level of salmon sperm. He had accomplished his work in the small laboratory in which his colleague the professor of chemistry, Piccard, gave practical courses to medical students in the Basel Museum. * It has been stated that Miescher, by isolating his "pure nuclein" from sperm heads, had discovered DNA. This requires some elaboration, since the reaction of xanthic bodies was located by Miescher in the other component, protamine, which was still an ill-defined compound when Miescher died of tuberculosis in a sanatorium in Davos (1895).

Miescher asked Piccard to define the substratum of the positive murexide reaction he had observed. Piccard [27] made succes-

* It was only in 1885 that the "Vesalianum" of Basel was built to house Miescher, as well as Bunge. The name of the Institute recalls the stay of Vesalius in Basel where he supervised the printing of the *De humani corporis fabrica libri septem* by the printer and humanist Oporinus.

References p. 263

sive extractions of salmon spermatozoa with hydrochloric acid of increasing strength. Final extracts, obtained with boiling hydrochloric acid, were found by Piccard to contain guanine and hypoxanthine. This was no mean achievement, considering the methods available at the time. Piccard made the pertinent remark that the murexide test obtained by Miescher with his "protamine" was probably due to a co-precipitation with the bases. Piccard suggested that guanine and hypoxanthine were not derived from the "protamine" but were present, along with it, in the sperm head. Piccard, nevertheless, did not refer to a presence of the purines (as they were called later) in the nucleic acid (as the "pure nuclein" of Miescher was called by Altmann in 1889).

But Miescher remained convinced of their presence in his "xanthin-alkaloid" (in fact the protein part of the heteroprotein). It must be recalled that, at the time, the xanthic bases were considered as protein constituents, a concept that was repudiated by Kossel in 1881 as stated in section 5 of this Chapter.

An interesting observation of Miescher was the slow diffusion and the great lability of his "pure nuclein" to which he assigned the empirical formula $C_{29}H_{49}N_9P_3O_{22}$ [28]. The ideas of Miescher on the nature of "impure nuclein" and "nuclein", served to confuse comparisons he performed on "impure nucleins" of different origins with respect to their content of sulphur, and justify the following statement of Fruton [23] (p. 186).

"About twenty years of work was required to sort out the confusion generated by these views".

As stated by Olby [29] (p. 97), Miescher, who aimed to found the staining properties of nuclei on a chemical foundation,

"not only failed to disperse the clouds of uncertainty but further confused the issue".

But the observations of Miescher and of Piccard established, in the circles interested in the pathology of gout and in uricogenesis, a bridge between the animal excretion of purines, and the presence in cell nuclei of the compounds which were more precisely characterized by Kossel.

7. Identification by Kossel of xanthic bases (later called alloxuric bases, purines) as components of nucleic acids

Clarification of the concepts introduced by Miescher started with the important work of Kossel. Kossel had studied medicine in Strassburg. In 1881 he was appointed as chief of the chemical section of the Institute of Physiology in Berlin. In 1895 he became professor of physiology in Marburg which he left in 1901 to succeed Kühne in Heidelberg. (On Kossel, see Eldbacher [30], Olby, [31], Jones [32].) The entry of Kossel into the nuclein field dates from the time when he was assistant to Hopper-Seyler in Strassburg.

In 1878 a paper appeared in which Nägeli and Loew [33] contested the assumption which had been presented by Hoppe-Seyler, of the presence of nuclein in yeast. Kossel [34], in 1879, in a paper on the nuclein of yeast, confirmed the presence, in it, of a characteristic substance distinguished by a high phosphorus content. We must remember that Miescher had considered that what he called "impure nuclein" (which we call nucleoprotein) was composed of a phosphorylated protein (corresponding to what we call the prosthetic group and to one of the compounds now called nucleic acids) and of what he called a xanthin-alkaloid, or protamine. We see that Miescher considered as the protein portion of the nucleoprotein (called by him impure nuclein) what was in fact the prosthetic group, which he called pure nuclein. Miescher considered that the xanthic bases were part of his xanthin-alkaloid, called by him protamine. That the xanthic bases are part of the prosthetic group (pure nuclein of Miescher) was clearly recognized by Kossel, a discovery which opened an entirely new pathway out of the mess introduced by Miescher.

As stated by Jones [35] (p. 4):

"Kossel observed that purine derivatives are always formed from nucleins when these substances are submitted to the action of hydrolytic agents, and he understood clearly that the bases originated from the "prosthetic group" (nucleic acid) and not from the protein of the nuclein. He first found hypoxanthine (1879) [36] and a trace of xanthine (1880) [37], then guanine (1882) [38]: (1883—1884) [39], and finally adenine (1886) [40]: (1888) [41]".

In this quotation, Jones, referring to the work of Kossel which

References p. 263

Plate 167. Albrecht Kossel.

started in 1879, uses terms introduced later; the term "purine" was introduced by Fischer in 1898, and the term "nucleic acid" by Altmann [61] in 1889. It was also recognized later that nucleic acids produce only two aminopurines (guanine and adenine). Oxypurines, such as hypoxanthine, are secondary oxidation products.

Jones [35] (p. 4) explains:

"As Kossel had not yet discovered adenine, it is but natural that he should have mistaken the substance for hypoxanthine, while the trace of xanthine which he sometimes found was probably a laboratory product formed from guanine in the execution of the Neubauer method [42] that was employed. Kossel's priority, however, is not in the least in question, for he himself found suitable methods of separating and identifying the bases (Schindler [43], Bruhns [44])".

Adenine was discovered by Kossel [45] in 1885. He isolated it from pancreas and he converted it into hypoxanthine by treatment with nitric acid.

To the compounds that were to be designated as "purines" by Fischer, Kossel gave the name of "alloxuric bodies" as he considered them to be composed of urea and of the alloxan nucleus, a name given by Liebig (1838) who considered this nucleus to be composed of oxalic acid and allantoin.

The importance of Kossel's work is recognizable in the definition he introduced in the characterization of the "pure nucleins" as containing those compounds he called alloxuric bases and differentiating from proteins the substances that had been called nucleic acids since 1889.

8. Pseudo-nucleins

This term, introduced by Hammarsten (see Fruton [23], p. 188), designated substances containing phosphorus and that were resistant to pepsin but not constituents of the cell nucleus. Such a substance had been isolated from milk by Lubavin [46] and another from egg-yolk by Bunge [47]. Kossel [48] showed that these compounds do not yield "alloxuric bases."

9. Dismissal of the notion of nucleins as phosphoproteins

The claim that "pure nucleins" are compounds of protein and metaphosphoric acid, in line with Miescher's ideas, was maintained by a series of authors (Liebermann [49]; Pohl [50]; Malfatti [51]) until Kossel [52] dismissed it in 1893.

We have noted above that the detection of "xanthic bases" in cell nuclei by Miescher and by Piccard had opened new vistas for the interpretation of the significance of those compounds, until then considered as involved in an excretion synthesis accomplishing the "detoxication" of the ammonia resulting from the deamination of amino acids. This opening of a new interpretation of the significance of purine bodies, which led to the concept of the biosynthesis of purines as a general property of cells, was highly reinforced by Kossel's work.

To quote Jones [35] (p. 4):

"It suggested in no uncertain way a chemical connexion between the cell nucleus and urinary uric acid, and was thus the foundation of many fruitful investigations in experimental medicine".

10. Structural formula of the purine nucleus and of its derivatives

Guanine, hypoxanthine, xanthine, uric acid (and the later discovered adenine) were recognized by Fischer [53] (1898) as derivatives of purine

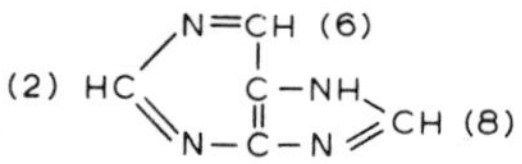

```
          N=CH (6)
         /   |
(2) HC       C-NH
        \\   ||     >CH (8)
          N-C-N  //
```

The detailed proof as the structure of each purine was epitomized by Fischer in 1907, in his *Untersuchungen in der Puringruppe* [54]. The starting point of these studies had been the isolation of alloxan and urea by Liebig and Wöhler, in their classical study of 1838, when they accomplished the oxidation of uric acid by nitric acid. Medicus, who was at the time assistant to Strecker in Würzburg, considered that this reaction limited the choice of possible structural formulas for uric acid to the following (see Jones [35], p. 15)

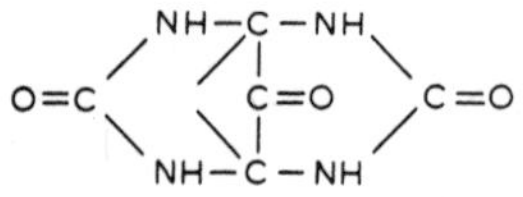

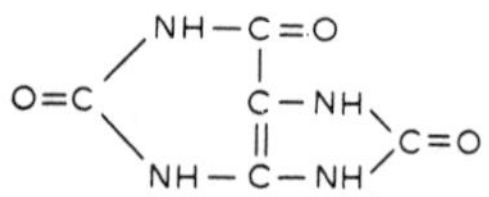

```
      NH-C-NH                      NH-C=O
     /  /|   \                    /    |
O=C     C=O   C=O            O=C       C-NH
     \  \|   /                    \    ||    >C=O
      NH-C-NH                      NH-C-NH
```

Medicus chose the version of the point of union between urea and alloxan represented on the right hand side. Medicus also proposed correct formulas for guanine, hypoxanthine, xanthine, as well as caffeine, which had been isolated from coffee by Pelletier and Caventou [56] in 1825. As stated by Jones [35],

"It seems that Medicus made a place for himself in history by his ability to guess the correct formula" (p. 15).

Experimental demonstration remained lacking until Behrend and Roosen [57] accomplished the synthesis of uric acid from urea and 4,5-dioxy-uracil (1889)

```
       NH—C=O
      /    |    ..............
O=C        C—:OH      H:HN
      \    ||  :    +   :    \
                              C=O
       NH—C—:OH      H:HN /
            ..............
```

The unravelling of the structural formula of uric acid became the key to the structure of the purine nucleus and of its derivatives. Fischer [58], in 1897, prepared the four purine bases from uric acid. Strecker [59], at the time professor in Christiania, converted amino-purines into oxy-purines by the action of nitrous acid. He converted guanine into xanthine. Sundwik [14], in 1912, converted, by direct reduction, uric acid into xanthine and hypoxanthine, which he called "sarkin", the name given by Strecker.

As a consequence of these works, and of the transformation into hypoxanthine, by Kossel [60], in 1886, of the adenine he had isolated from the hydrolytic products of pancreatic nucleic acid, the relation of the five purine compounds to one another and to purine can be represented as follows (see Jones [35], p. 14).

```
  H (2)        NH2           H            OH            OH            H
 /            /             /            /             /             /
P—H (6)      P—OH          P—NH2        P—OH          P—OH          P—OH
 \            \             \            \             \             \
  H (8)        H             H            OH            H             H
purine       guanine       adenine      uric acid     xanthine      hypoxanthine
             (2-amino-6-   (6-amino-    (2-6-8-       (2-6-dioxy-   (6-oxy-
             -oxy-purine)  -purine)     -trioxy-      purine)       purine)
                                        purine)
```

11. The hydrolytic products of thymus nucleic acid

Concentrating on the nucleic acid (a term introduced by Altmann [61] to designate the prosthetic group of nucleoproteins) obtained from thymus, Kossel and Neumann [62,63] were surprised to find among its hydrolytic products, only one purine, adenine. Previously, examining nucleic acids of different origins, Kossel [64] had found them different with respect to their content in the different purines.

Would a large number of different nucleic acids exist in the living world? This Kossel thought improbable, and he was led to suppose that there are four kinds of nucleic acid, each containing a specific puric base, and that variable mixtures of the four types existed in different kinds of organisms. We find here an expression of the old idea of the chemical unity of organisms with a superimposed diversity resulting from the differences in relative amounts of the same components. Kossel and Neumann eventually obtained two purines: adenine and guanine and, by the application of more powerful hydrolytic agents that destroyed the purines,obtained thymine [62] (1893) and cytosine [63] (1894). Thymine was easily isolated, and received from Kossel and Neumann the empirical formula $C_5H_6N_2O_2$.

Cytosine gave more difficulty. Its correct formula, $C_4H_5N_3O$ was recognized later by Kossel and Steudel [65] who obtained it by the action of nitric acid on uracil. In the same work, by strong hydrolysis, Kossel and Neumann obtained levulinic acid which they recognized as derived in the analytical process from a hexose.

12. The hydrolytic products of yeast nucleic acid

Yeast nucleic acid was first isolated by Altmann [61] (1889), but he did not study its chemistry. Kossel [64] isolated from it guanine and adenine, and recognized the presence of a pentose. Later, in Kossel's laboratory, Ascoli [66] isolated uracil from yeast nucleic acid, and Kossel and Steudel [65] obtained cytosine in 1903.

13. An alleged chemical difference between plants and animals, concerning the hydrolytic products of nucleic acid

Nucleic acids were recognized by Kossel as dehydrolysed products of phosphoric acid, a carbohydrate and four nitrogen ring compounds (see Jones [35]). For a long time, the notion persisted in comparative biochemistry of the specific nature of plant and animal nucleic acids based on the following pattern:

Plant nucleic acid (yeast nucleic acid)	*Animal nucleic acid (thymus nucleic acid)*
Phosphoric acid	Phosphoric acid
Guanine	Guanine
Adenine	Adenine
Cytosine	Cytosine
Uracil	*Thymine*
Pentose	*Hexose*

14. Pyrimidine derivatives of nucleic acids

As stated above, Kossel and his coworkers discovered cytosine, thymine and uracil.

The latter had already been mentioned by Behrend as a hypothetical substance from which the pyrimidines were derived (see Jones [35] p. 17). Jones has retraced the steps that led to recognition of the structural formulas of the following pyrimidines, later confirmed through synthesis by a number of organic chemists. The possible derivation of the pyrimidines from the purines in the course of preparation has been disposed of. (Literature in Jones [35], p. 21.)

```
           (1)  (6)
           NH—C=O
          /    |
(2) O=C        CH (5)       thymine
          \    ||
           NH—CH (4)
           (3)

           N══C·NH2
          /    |
      O=C      CH           cytosine
          \    ||
           NH—CH

           NH—C=O
          /    |
      O=C      CH           uracil
          \    ||
           NH—CH
```

References p. 263

Plate 168. J. Horbaczewski.

15. The concept of an essential difference in the process of uricogenesis in birds and in mammals

The work of Kossel on nucleic acid suggested a metabolic relation between excreted uric acid and the nucleic acids of cell nuclei. A biosynthetic formation of uric acid in birds appeared as substantiated by the demonstration of the formation, in birds, of uric acid from amino acids (see Section 1 of this Chapter) as well as by the experiments of Minkowski [5]. On the other hand, a connection between nucleic acids and uric acid had been investigated since the work of Stadthagen [67], as far back as 1887. Stadthagen did not observe any increase of uric acid excretion after feeding dogs with nucleic acids or guanine, and his negative results were confirmed by Gumlich [68] in 1894. It was not realized at the time that dogs excrete allantoin instead of uric acid.

In his classical work on geese, Minkowski [5] had observed that the birds with extirpated livers excrete ammonium salts, but he also noted that uric acid never, even in hepatectomized geese, completely disappeared from the excreta. This observation of Minkowski called the attention of von Mach [69], who injected hypoxanthine into hepatectomized geese, subcutaneously, and found an increase of uric acid excretion.

As noted by Jones [35] (p. 65):

"The ability of the animal organism to form uric acid from a purine precursor was thus shown for the first time, but the origin of uric acid was not traced to nucleic acid".

This was done by Horbaczewski. He had realized in 1882 [70] the first complete synthesis in vitro of uric acid from glycine and urea. He therefore gave glycine to dogs, but obtained no uric acid excretion. This was confirmed by Weiss [71]. Steudel [72], after giving uracil and other pyrimidine derivatives to dogs, observed no uric acid excretion. * Horbaczewski was the first to observe in man a formation of uric acid from nucleic acids or from food rich in nucleic acids.

The interpretation of Horbaczewski was that nucleic acids

* It was not, before 1907, generally known that mammals, except man or anthropoid apes, excrete allantoin instead of uric acid.

References p. 263

induced leucocytosis and that it was from dead leucocytes that the increase of uric acid excretion orginated [74]. The motivation of Horbaczewski's experiments are found in medical notions and in particular in the fact that patients with leukemia excreted large amounts of uric acid and presented a high leucocyte count. Though the leucocytosis theory has long been dead, the increase of uric acid excretion in man after the ingestion of nucleic acids or of food rich in nucleic acid (such as thymus) has been amply confirmed. (Literature in Jones [35], p. 66.)

Besides his positive observation on the effect of feeding nucleic acids to man, Horbaczewski described experiments on putrefying calf spleen, in which he claimed that uric acid is formed when air it passed through the mixture, whereas in its absence, xanthine and hypoxanthine are formed. Though not easy to interpret, these experiments of Horbaczewski gave the first hint of the method of production of uric acid in man by what has since been called "oxidative formation". As he had convinced himself that amino acids do not increase the excretion of uric acid in man, although nucleic acids do, Horbaczewski propounded the concept of the entirely oxidative origin, "in mammals" (in fact in man), of uric acid from hypoxanthine or xanthine (mammalian pattern), to the exclusion of biosynthesis, as was recognized to occur in birds (bird pattern).

16. Physiological conversion of oxy-purines to uric acid

Horbaczewski's work, in the frame of a scientific community highly interested in the uricosaemia of gouty patients and in the high uric acid excretion of leukaemics, was highly appreciated. It stimulated the study of metabolic relations between the excretion of uric acid and the metabolism of purines. Even before Horbaczewski's publication, von Mach [69], demonstrated that, after hepatectomy, a bird is still able to convert subcutaneously injected hypoxanthine into uric acid. This was interpreted as in favour of the existence, in birds, of both "oxidative" and "synthetic" uricogenesis. Nevertheless, von Mach was the first to observe in vivo a conversion of an oxy-purine into uric acid. This point was the subject of further investigations of Minkowski, after the ideas of Horbaczewski on the "oxidative path" were publicized. As we have stated, Horbaczewski had observed in man a production of uric acid from nucleic acids or from food rich in nucleic acids. This positive result was important, as no such effect could be obtained with free purines in man. Kossel [75], on giving adenine to dogs, had found that it created severe disturbances and was to some extent unchanged. Minkowski [76] showed in 1898 that no such toxic effect was observed in dog or man after injecting hypoxanthine, and he observed an increased excretion of uric acid in man.

In contrast with oxy-purines, free amino-purines, contrary to their fate when combined in nuclein, remained incompletely transformed into uric acid in the human organism.

That the oxy-purines xanthine and hypoxanthine were oxidized to uric acid by aqueous extracts of ox liver and of ox spleen, when a current of air was passed through the digesting mixture, was observed by Spitzer [77] (1899). The enzyme involved was called xanthine oxidase by Burian [78]. At the time of Spitzer's observations, it was known from the work of Schindler [43] (1889) that amino-purines are converted into oxy-purines by putrefactive bacteria.

After a rather reckless controversy (literature in Jones [35], pp. 71–74) it was recognized that enzymatic agents liberated the amino-purines guanine and adenine from nucleic acid (an action that was explained later when the chemistry of nucleic acids became better known), and that the amino-purines are deaminated in the presence of guanase and adenase. The resulting oxy-purines, xanthine and hypoxanthine, are oxidized to uric acid in the presence of xanthine oxidase.

17. The alleged synthetic formation of uric acid by way of urea

The current view during the last decades of the 19th century was in favour of an essential difference between man and birds. It was considered that, in the human organism, uric acid resulted from an oxidative transformation of purines, whereas in birds it was the result of a biosynthetic process considered as an extension of the pathway of the deamination of amino acids.

References p. 263

No relation was considered to exist between the biosynthesis of the purines of nucleic acids and the excretion of uric acid, except in what concerned the process of purine oxidation.

This opposition between the two clear-cut metabolic processes was shaken by the conclusion stated above that uric acid, in birds, is produced by both, admittedly to a smaller extent by the "oxidative" pathway than by the "synthetic" one. Since the ability of birds to form uric acid by two metabolic paths was widely accepted, it was thought logical to consider, as stated by Wiener [79]:

"dass auch bei Säugetieren beide Entstehungsweisen der Harnsäure in Betracht kommen".

That uric acid can, in man, be obtained by the synthetic path had been claimed, in a study on gouty deposits, by Freudweiler [80]. This author suggested that the differences between man and bird are only quantitative. He considered that the biosynthetic pathway of uric acid formation had been demonstrated by Nencki, Pavlov and Zaleski [81] to take place in mammals. This is a result of misreading their paper and confusing uric acid (Harnsäure) with urea (Harnstoff). Furthermore, dogs do not excrete uric acid.

The persistent claim for the existence of a biosynthetic pathway for uric acid distinct from the pathway of the biosynthesis of nucleic acids (at the time completely unknown and without analogy in organic synthesis) was dependent on an epistemological aspect, the participation of urea in the schemes of synthesis of uric acid in vitro. For instance, Behrend and Roosen [82] (1888) synthesized uric acid by warming a mixture of 4,5-dihydroxy uracil and urea with concentrated sulphuric acid.

Horbaczewski [70] had himself developed a synthesis of uric acid in vitro from glycine and urea (1882) and later [83] (1887) another synthesis from urea and trichlorlactamide. But though considered as obvious from the point of view of the chemical nature of uric acid, its derivation from urea was first proposed in the form of a biosynthetic pathway by Wiener, who worked in the Laboratory of Pharmacology of the Deutsche Universität in Prague. Wiener [79] was convinced of the existence of a synthetic pathway of uric acid formation, distinct from the oxidative process starting from nucleic acid purines, by an observation he made using the isolated ox liver. Having observed a production of uric acid by such an isolated organ, he recognized an increase of this production after adding to the perfusion fluid an alcoholic extract of another liver. Such an extract, giving no precipitate on the addition of ammoniacal silver, was certainly deprived of purines. It had been stated by Meyer [84], in a dissertation presented at the Königsberg University in 1877, that feeding urea increases the excretion of uric acid in birds. Wiener concluded from his experiments that the increased uricogenesis, as he believed, from urea, is not the result of a diuretic effect. Aiming, according to current ideas of the period, to identify substances able to combine with urea to produce uric acid, Wiener tried, in his hens, a number of compounds: glycerol, propionic acid, lactic acid, ethylene lactic acid, pyruvic acid, malonic acid, tartronic acid, mesoxalic acid, butyric acid, α-oxybutyric acid, β-oxybutyric acid, succinic acid and malic acid. His results he interpreted in terms of a metabolic theory leading, by oxidation from monocarboxylic acids of the alipathic series to acids with three carbons. If such an acid is propionic acid, it is further metabolized into water and carbonic acid. If the product is a ketonic or an oxy-acid, a dicarboxylic acid results. According to Wiener:

"These are not further metabolized but completely converted into uric acid, when enough nitrogenous substance is available" (translation by the author).

Wiener found no production of uric acid in the dog. In man, he explored the effect of feeding lactic, malonic and dialuric acids (as he could not obtain enough tartronic acid, he used its diureide). To facilitate the synthesis of uric acid, Wiener gave, along with the acids he tried, a supplement of urea. He also performed control experiments with sodium acetate. The increased production of uric acid observed by feeding tricarboxylic or dicarboxylic acids or precursors of these was, as other authors have remarked [85], very limited.

On the other hand, when considering the observations cited by Wiener to substantiate his contention of an increased excretion of uric acid in fasting hens after feeding dialuric acid, we see that he, for instance, considered the mean normal daily excretion as amounting to 0.5377 g of uric acid compared with the mean value of 0.6082 for the days when dialuric acid is given. But his values show that such an alleged increase of 0.06—0.07 g uric acid is within the range of the variations of the values from which the

References p. 263

value corresponding to normal days is derived. In fact, none of the differences claimed by Wiener is significant (R.B. Fischer [86], in collaboration with R.A. Fisher). The theory proposed by Wiener is based on a combination of urea with tartronic acid, forming the diureide of this tricarboxylic acid, a second molecule of urea being involved in the formation of uric acid.

$$\underset{\text{urea}}{\begin{array}{c} NH_2 \\ | \\ CO \\ | \\ NH_2 \end{array}} + \underset{\text{tartronic acid}}{\begin{array}{c} COOH \\ | \\ CHOH \\ | \\ COOH \end{array}} = \underset{\text{dialuric acid}}{\begin{array}{ll} NH-CO & \\ |\quad\;\; | & \\ CO\quad CHOH & \\ |\quad\;\; | & \\ NH-CO & \end{array}} + 2\,H_2O$$

$$\underset{\text{dialuric acid}}{\begin{array}{ll} NH-CO & \\ |\quad\;\; | & \\ CO\quad CHOH & \\ |\quad\;\; | & \\ NH-CO & \end{array}} + \begin{array}{l} H_2N \diagdown \\ \qquad\quad CO \\ H_2N \diagup \end{array} = \underset{\text{uric acid}}{\begin{array}{l} NH-CO \\ |\quad\;\; | \\ CO\quad C-NH\diagdown \\ |\quad\;\; \| \qquad\quad CO \\ NH-C-NH\diagup \end{array}} + 2\,H_2O$$

18. Catabolism of purines (uricolysis)

While not belonging to biosynthetic studies, uricolysis, which was not treated in Part III of this *History* must be briefly treated here.

We have noted (Chapter 47) the surprise of Fourcroy when in 1793, he did not find uric acid in a urinary calculus of a horse. Was uric acid, he asked, only present in man? In fact, at that time (1790), allantoin had already been isolated from allantoic fluid by Vauquelin who had called it amniotic acid or allantoic acid. In their classical paper of 1838, Liebig and Wöhler, having obtained it by the oxidation of uric acid, called it allantoin ($C_4N_4H_6O_3$). Allantoin was found in calf's urine by Wöhler [87] in 1849. That the presence of allantoin in the urine of mammals other than man had not become common knowledge thereafter is shown by the conclusion reached by Frerichs and Städeler [88]. Having detected allantoin in the urine of dogs with artificial dyspnoea, they considered its appearance as resulting from an insufficient supply of air. This shows that they (as well as the editors of the journal) were not aware of the normal presence of allantoin in the urine of dogs on the one hand, and, on the other hand, that they could not have conceived of a production of allantoin from uric acid by oxidation.

The occurrence of allantoin in the normal urine of dogs, cats and rabbits had been first recognized by Meissner [89] in 1868. But, as we shall see, the notion was only generally accepted after 1907. The lack of interest in comparative biochemistry and the over-emphasis, due to the interest in gout, on the excretion of uric acid by man, maintained the erroneous notion of the excretion of uric acid by mammals.

What is of interest to us in the present context is the notion, now acquired, that uric acid, as well as allantoin are catabolic products of the purines adenine and guanine, components of nucleic acids, and not products of another biosynthetic pathway going, for instance, through the intermediary step of urea as was long believed.

One of the obstacles to the recognition of the uricolytic pathway, and of the derivation of allantoin from uric acid in mammals except man and anthropoid apes, was that it was believed since Wöhler and Liebig that uric acid introduced as such into the organism of a mammal was converted into urea. Even if there were an increased excretion of uric acid in the urine, this was considered as a result of leukocytosis and of the degradation of leukocytes, a concept derived from the high excretion of uric acid in leukemia (see Folin [90]). This theory remained current for a long time, as a consequence of the high authority of Wöhler as a chemist. That uric acid introduced into man was excreted as such, and not in the form of urea, was first claimed by Soetheer and Ibrahim [91] in 1902.

With respect to the interpretation of the presence of allantoin in the urine of mammals, the urea theory was still invoked by Minkowski [92] in 1898. Finding that feeding dogs with calf thymus increased their excretion of allantoin, he concluded that allantoin was more slowly transformed into urea in dogs than in man. This reasoning was accepted by several other authors of the period (Cohn [93]; Salkowski [94], Poduschka [95], Mendel and Brown [96], etc.). That catabolized uric acid was, in the general case of mammals, quantitatively excreted as allantoin was first stated by Wiechowsky [97] in 1907.

Wiechowski observed that uric acid was quantitatively converted into allantoin by extracts of liver of animals excreting allantoin but not by extracts of human tissues. A number of authors (Winternitz and Jones [98], Miller and Jones [99], Schittenhelm [100]) agreed with Wiechowski's conclusion, which passed into general acceptance. The mechanism of conversion of uric acid into allantoin remained mysterious for some time. Chemical models had been studied by chemists (Behrend [101], Biltz and Max [102], Biltz and Schauder [103]) who obtained by oxidation several degradation products of uric acid. These were tried on animals (literature in Brünig et al. [104]) but proved not to be possible precursors of allantoin.

Uricase, the enzyme catalysing the transformation of uric acid to allantoin, i.e. opening the purine ring, was named in 1909 by Battelli and Stern [105]. The mechanism of action of uricase was elucidated by the researches of Felix, Scheel and Schuler [106], of Schuler and Reindel [107] and of Fosse, De Graeve and Thomas [108]. They identified an intermediary product, oxyacetylene-diureine-carbonic acid. Uricase catalyses an oxy-hydratation and a decarboxylation.

In 1925, Przylecki [109] observed a degradation of allantoin into urea in the presence of amphibian livers. It was therefore concluded that this process requires two enzymes. The name allantoinase was first given to the whole system by Nemec [110]. At the suggestion of Fosse and Brunel [111], the name was reserved for the hydratase which by fixation of a molecule of water onto allantoin, transforms it into allantoic acid by opening the imidazole ring. The name allantoicase was independently given by Krebs and Weil [112] and by Brunel [113] to the hydrolase catalysing the hydrolysis of allantoic acid with the formation of one molecule of glyoxylic acid and two molecules of urea.

Urease is the well-known amidase showing a very strict specificity towards urea, which is hydrolysed at the C—N bond. The whole system of uricolysis, or of the catabolic pathway of purines, from uric acid to ammonia is frequently found in plants (Fosse [114]). Its importance in the metabolism of germinating seeds has been emphasized by the school of Fosse in Paris (Echevin and Brunel [115], etc.). In mushrooms, the same pattern occurs and allows for the utilization of ammonia derived from urea, allantoic acid, allantoin or uric acid for building the mycelium (Brunel [113]). A study of enzyme distribution in animals (literature in Florkin [116]) leads to the conclusion that if the enzyme uricase does not exist in the tissues, uric acid is the end-product of purine catabolism. This is true for man and anthropoid apes, birds, terrestrial reptiles, cyclostomes and many insects. If the enzyme uricase uric acid

↓ *uricase*

allantoine

↓ *allantoinase*

allantoic acid

↓ *allantoicase*

glyoxylic acid + 2 $O{=}C(NH_2)_2$ urea

↓ *urease*

$2\,NH_3 + CO_2$

ammonia

occurs in the tissues as it does in many mammals (except man and the anthropoid apes) the end-product is allantoin, if this compound is not further catabolized.

Amphibian livers have been found, as stated above, to decompose uric acid to urea, a process involving uricase, allantoinase and allantoicase [117—119].

Some teleost fishes convert uric acid to allantoin and further to allantoic acid, whereas most fishes, amphibia and fresh-water pelecypods, containing allantoicase, degrade uric acid to glyoxylic acid and urea (Brunel [120]). The complete path of purinolysis, leading to CO_2 and ammonia is found in the lower animal forms, such as marine pelecypods, crustacea and sipunculids, equipped with uricase, allantoicase and urease [121].

References p. 263

In the frame of the evolutionary approach, it appears from the knowledge of the distribution of uricolytic enzymes and of the physiological data available that the complete chain leading to ammonia formation, occurring in bacteria and in plants, also exists in marine invertebrates such as *Sipunculus*, mussel or marine crustaceans.

"By the loss of a terminal link, the enzymatic chain leads next only to the formation of urea, as it is seen in the fresh water mussel (*Anodonta*), fishes and amphibians. A further shortening of the chain by one or two links, leading to the formation of allantoin or of uric acid, is found in animals which have become more highly evolved in the sense of attaining greater independence or by way of adaptation. This is the case in oligochetes, leeches, insects, reptiles, birds and mammals.

"A study of the zoological distribution of enzymes of the uricolytic system, thus, shows that the evolution of the purine metabolism is attained by a process of simplification through the dropping off of terminal links in the uricolytic enzyme chain. The evolution presents itself as a regressive biochemical orthogenesis through a process of enzymapheresis". ([122] Translation by S. Morgulis, p. 49.)

19. Theory of allantoin biosynthesis from urea

During the first years of the present century, before the derivation of allantoin from uric acid was accepted but after the experiments of Wiechowski (1907), the excretion of allantoin by mammals known since Meissner's demonstration (1868) was still considered, as well as the excretion of uric acid by birds, as the result of a biosynthetic process de novo and distinct from the (unknown at the time) biosynthetic production of the purines of nucleic acids. This concept, and the recourse of the pathway described by Wiener to account for its implementation, is found in a paper of Eppinger [123], published in 1905. Eppinger recognized on the one hand the existence of an oxidative process of allantoin derivation from purines, but he claimed the accomplishment of a synthesis of that compound de novo. Eppinger started from the formation of allantoin in vitro by heating a mixture of glyoxylic acid and urea, or from mesoxalic acid and urea. Feeding glycolyldiurea to mammals he obtained an increased allantoin excretion although no such effect was obtained in man. These results reinforced the belief in the validity of Wiener's theory, which was to remain current for the three following decades. (It was still accepted in the book of Jones [35] on nucleic acids, published in 1920.)

20. Opposition to Wiener's theory

The fact that it was not repudiated until the 40's gives an idea of the ingrained state of the theory of the formation of uric acid de novo from the ammonia resulting from amino acid deamination by way of an intermediary formation of urea as was supposed to take place with a particular intensity in the livers of birds. Opposition towards this viewpoint nevertheless found an expression long before the theory disappeared, mainly by the recognition that the ornithine cycle was lacking in birds, an essential discovery as it excluded without doubt a participation of urea in the process. The opposition finds, for instance, an expression in the Habilitationsschrift of Pfeiffer [124]. A collaborator of Quincke in the laboratory of clinical research of Kiel, Pfeiffer accomplished his work partly in the Institute of Physiological Chemistry of Strassburg.

Pfeiffer pointed to the wide scattering of the results obtained by Wiener. Feeding man or monkeys with malonamide, tartronamide or tartronic acid, he observed no appreciable increase of uric acid excretion. As his motivation was studies on gout, he concluded that, in that disease, the increase of uric acid concentration in blood could not be related to an increased biosynthesis (meaning a biosynthesis de novo). This is one more testimony of the complete lack of taking purine biosynthesis into consideration. Pfeiffer suggested a lower purine oxidation or a reduced kidney excretion.

But, in experiments on birds-liver "brei", the theory of Wiener retained support for a long period by the publication of Ascoli and his group (Ascoli [125]; Ascoli and Izar [126], Bezzola, Preti and Izar [127], Izar [128], Preti [129]). Attempts to repeat these observations have nevertheless consistently failed (Spiers [130]; Calvery [131]). Again, in 1928 and 1930, arguments in support of Wiener's theory were presented by Pupilli [132,133] who claimed that in the hen, urea is conjugated with the compounds used by Wiener, to yield uric acid, and he claimed to have obtained the same results with hen's liver powder. An objection to these results was that the method used by Pupilli for estimating

References p. 263

uric acid (Ganassini's method) was not specific.

On the other hand, Clementi [34], who had maintained for years that birds do not synthesize urea (it was a part of the background of Clementi's law) showed in 1930 that injected urea can be quantitatively recovered from the excreta of birds whether or not pyruvate or malonate is administered simultaneously. These conclusions were, during the next year, confirmed by Torrisi and Torrisi [135] and by Russo and Cuscuna [136]. In addition, Torrisi and Torrisi showed that the administration of pyruvate and malonate had no influence on the excretion of uric acid by the hen.

21. Concept of a pathway of purine synthesis from ammonia without participation of urea

Still accepted in Jones's book [35] on nucleic acids (1920), the concept of the participation of urea as an intermediate in uricogenesis in birds had nevertheless been attacked from another viewpoint through the observations of Kowaleski and Salaskin [137] who, rather than administering the precursors considered by Wiener, showed that when a goose liver is perfused with blood, uric acid appears in the blood. If they added ammonium lactate to the blood the concentration of uric acid was increased. In view of the consistent failures recorded in the literature to obtain urea synthesis in preparations of isolated bird liver, the observations of Kowaleski and Salakin were against the theory of Wiener and they suggested the existence of another biosynthetic pathway not involving urea, and through which ammonia acted as a precursor of the purine nucleus.

In 1933, Schuler and Reindel [138] published experiments on tissue slices of pigeon's organs, accomplished in the chemical laboratory of the medical clinic of Erlangen University. They concluded that kidney, as well as liver, may take part in the biosynthesis of the purine ring. These experiments excluded a participation of urea. In a second paper, also published in 1933, Schuler and Reindel [139] claimed that the synthesis of uric acid in pigeons implies a precursor which is converted into uric acid in the kidney and which may be formed in the liver as well as the kidney. In another paper of the same year [140], they extended the concept to the process of uric acid synthesis in the hen and in the goose. Simultaneously, similar views were proposed in a paper of Benzinger and Krebs [141]. To quote Krebs [142]:

"After the successful work on the urea synthesis, I began at once to tackle the problem of uric acid synthesis in birds, a process physiologically analogous to urea synthesis, being the main process in birds by which ammonia is detoxicated and neutralized. This work was undertaken together with another medical student, T.H. Benzinger. We used pigeons for the first experiment because it was the most readily available bird for experimental purposes. To our great surprise, the pigeon liver slices failed to form uric acid whilst chicken liver slices when tested gave clear-cut positive results. Since the pigeon is no exception to the rule that birds excrete their excess nitrogen mainly in the form of uric acid, we searched other organs for their capacity to form uric acid, and in the course of these experiments we discovered that when slices of liver and kidney were incubated together, or when the medium in which liver slices had been shaken is subsequently incubated with kidney slices, uric acid is formed. Kidney slices alone did not form much uric acid. These findings raised the question of the product formed by the liver and subsequently converted to uric acid by the kidney" (p. 22).

22. Synthesis of hypoxanthine from ammonia and unknown carbon precursors

Krebs, having left Freiburg University at the advent of Nazism, was obliged to interrupt his researches on the uric acid synthesis in birds. In England he first worked in Cambridge, and afterwards in the Department of Pharmacology of Sheffield University. In 1936 there appeared a paper of Edson, Krebs and Model [143], the conclusion of which was that

"(*1*) In pigeons two tissues are necessary for uric acid synthesis. The primary step, the binding of ammonia with some uncertain source of carbon, occurs in the liver where hypoxanthine is formed.

(*2*) The final conversion into uric acid, which takes place in the kidney (or pancreas) is an oxidation catalyzed by xanthine oxidase.

(*3*) There is evidence suggesting that hypoxanthine is also an intermediate in other birds."

In 1937, Reindel and Schuler [144] published the results of researches accomplished in the Medical Clinic of Erlangen and in the 2nd medical clinic of Munich. They clearly referred to these studies as related

"Zur Bildung von Nucleinsäuren notwendigen Aufbau des Purinringes".

References p. 263

Adding *l*-alanine to slices of pigeon liver, they detected the production of a single purine, xanthine. This remained unexplained, as Orström, Orström and Krebs [145] confirmed that the purine base synthesized by pigeon liver requires 1 mole O_2 for its quantitative conversion into uric acid and is therefore hypoxanthine.

23. Early history of the sources of the carbon atoms of the purine ring

It was only after World War II, as we shall see (Part V), that the sources of the atoms of the purine ring were finally identified. These studies showed that

"The purine skeleton is built up on the nitrogen atom of ribosylamine 5-phosphate. Substrates utilized in this process include glutamine, glycine, aspartic acid, carbon dioxide and a one-carbon fragment, equivalent to formaldehyde which is supplied in the form of formyltetrahydrofolic acid" (Krebs [142], p. 22).

Possible precursors of the carbon chain of purines had been investigated during the period before World War II. (Reviews: Rose [146], Christman [147].) Tartronic acid, suggested by Wiener, was, as stated above, discarded.

Many other possible precursors, proteins, fats or carbohydrates have been tested as possible precursors of the purine nucleus. In general, these experiments in vivo were performed on fasted animals. The administration of proteins increased the excretion of purine derivatives [148—152]. Under conditions in which ammonia did not increase uric acid excretion, alanine, glycine, asparagine and glutamine produced an increase of uric acid excretion in man [153—156].

Attention was naturally directed to the deaminated derivatives of these amino acids. R.B. Fischer [86], in the laboratory of Peters in Oxford, followed this trend and claimed that L-lactate (not D-lactate) may increase uric acid excretion in the bird. On the contrary, Gibson and Doisy [157] found that, in man, lactate decreased the excretion, as did glycollic acid, whereas alanine and pyruvate increased it.

The experiments in vivo that have just been related were recognized later as highly significant because they showed the importance of the carbon chain of compounds such as glycine, aspartic acid and glutamic acid in purine biosynthesis. But, unexplained as they were at the time, such observations, though later proved relevant, were criticized on the basis of a supposed "depletion of body stores". The concept of the gradual increase, in fasting animals (generally used in these experiments), of stores of uric acid in blood and tissues was introduced by Lennox [158]. These stores were considered as released under the influence of proteins or amino acids. Nevertheless, it should have been noted that, in experiments accomplished in vitro on pigeon liver slices, Örström, Örström and Krebs [145], in Sheffield, had demonstrated that asparagine, glutamine, pyruvate or oxaloacetate accelerated the rate of hypoxanthine formation. But the authors themselves pointed out that the stimulatory effect of these acids did not necessarily indicate a participation of their carbon atoms in purine synthesis. In such context, the theory which became current between the two World Wars was the theory of Ackroydt and Hopkins, according to which the precursors of the carbon atoms of the purine ring were arginine and histidine.

24. Arginine and histidine

Ackroydt and Hopkins [159], postulating that the imidazole nucleus of allantoin could be derived from arginine or histidine, had claimed (1916) that the removal of arginine or histidine from the diet of the young rat diminishes the excretion of allantoin by 40—50 per cent. This was the basis of the theory that arginine and histidine were precursors of the purine nucleus. The evidence for this theory nevertheless remained slender and limited to one species, the rat. For instance, no demonstration of an effect, in human adults, of comparable diets differing in their contents of arginine or histidine could be accomplished by Lewis and Doisy [160].

As placental proteins were considered to contain high proportions of arginine, the idea occurred to Harding and Young [161] that this amino acid was an important precursor in the biosyntheses of the embryo. They therefore fed to puppies meat proteins or placental proteins. With the placental diet, rich in arginine, more allantoin was excreted than with the meat diet, an observation considered by the authors as showing that arginine is the mother substance of the purine ring. Rose [162] criticized such a

References p. 263

conclusion, believing that the results could be explained by the difference of the contents of the diets in preformed purines.

Repeating the experiments of Ackroydt and Hopkins on young rats, Rose and Cook [163] concluded that under ordinary conditions of diet, histidine should be recognized as the mother substance of allantoin. They also concluded that arginine cannot replace histidine. But, in infants of 4—8 months, György and Thannhauser [164], adding histidine to diets of milk and of casein hydrolysates from which most of the arginine and histidine had been removed, failed to observe any influence on the excretion of uric acid.

On the other hand, Crandall and Young [165], who fed puppies on synthetic diets containing a single protein, observed that haemoglobin, rich in histidine, caused the largest excretion of purine derivatives. Though results obtained by the authors mentioned, and by others who gave diets high ot low in arginine and histidine, have failed to agree or to be conclusive (Abderhalden and Einbeck [166]; Abderhalden, Einbeck and Schmid [167]; Young, Conway and Crandall [168]), the influence of the discovery, by Krebs and Henseleit, of the ornithine cycle continued to suggest a participation of arginine in the biosynthesis of the ureide group of uric acid.

As we shall see, one of the first major advances accomplished through the use of isotopic tracers after World War II was the repudiation of the alleged participation of arginine or histidine as precursors of the purine ring.

25. Pyrimidines considered as precursors of purines

We have referred in Section 10 of this Chapter to a synthesis of uric acid from a pyrimidine. In 1930, Cerecedo [169] considered the possibility of a biosynthesis of the purine ring from pyrimidines but the experimentation showed that pyrimidines, in the mammalian organism, are oxidized to urea.

REFERENCES

1 E. Salkowski, Z. Physiol. Chem., 1 (1877) 1.
2 C.O. Cech, Ber. Dtsch. Chem. Ges., 10 (1877) 1461.
3 W. von Kniriem, Z. Biol., 13 (1877) 36.
4 W. von Schroeder, Z. Physiol. Chem., 2 (1878) 228.
5 O. Minkowski, Arch. Exp. Pathol. Pharmacol., 21 (1886) 41; 31 (1893) 214.
6 Sir R. Peters, in J. Needham (Ed.), The Chemistry of Life, Cambridge, 1970, p. 194.
7 C. Rutz, The Garrods, Zürich, 1970.
8 A.B. Garrod, The Nature and Treatment of Gout and Rheumatic Gout, London, 1860.
9 F. Wöhler and J. Liebig, Ann.Chem., 26 (1838) 340.
10 F. Lieben, Geschichte der physiologischen Chemie, Leipzig and Vienna, 1935, pp. 576—583.
11 A. Marcet, An Essay on the Chemical History and Medical Treatment of Calculous Disorders, London, 1817.
12 J.J. von Scherer, Ann. Chem., 73 (1850) 328.
13 A. Strecker, Ann. Chem., 108 (1858) 129.
14 E.E. Sundwik, Z. Physiol. Chem., 76 (1911—1912) 486.
15 B. Unger, Ann. Chem., 108 (1858) 129.
16 R. Virchow, Arch. Pathol. Anat. Physiol., 35 (1866) 358.
17 N.G. Coley, Ambix, 18 (1971) 69.
18 G. Brugnatelli, Ann. Chim. Phys., 8 (1818) 201.
19 W. Prout, Phil. Trans. Roy. Soc., 108 (1818) 420.
20 W. Prout, Ann. Phil., 14 (1819) 363.
21 G. Bird and R.H. Brett, London Med. Gazette, 14 (1834) 751.
22 J.R. Partington, A History of Chemistry, vol. 4, London, 1964.
23 J.S. Fruton, Molecules and Life, Historical Essays on the interplay of Chemistry and Biology, New York, 1972.
24 R. Olby, in C. Gillispie (Ed.), Dict. Sci. Biogr., vol. 9, New York, 1974, p. 380.
25 F. Miescher, Med. Chem. Unters. (first name of the journal which became the Z. Physiol. Chem.), (1871) 441.
26 F. Miescher, Ber. Dtsch. Chem. Ges., 7 (1874) 376.
27 J. Piccard, Ber. Dtsch. Chem. Ges., 7 (1874) 1714.
28 F. Miescher, Verhandl. Naturforsch. Ges. Basel, 6 (1874) 138.
29 R. Olby, The Path to the Double Helix, London, 1974.
30 S. Edlbacher, Z. Physiol. Chem., 177 (1928) 1.
31 R. Olby, C. Gillispie (Ed.), Dict. Sci. Biogr., vol. 7, New York, 1973, p. 466.
32 M.E. Jones, Yale J. Biol. Med., 26 (1953) 80.
33 C. Nägeli and O. Loew, Ann. Chem., 193 (1878) 322.
34 A. Kossel, Z. Physiol. Chem., 3 (1879) 284.
35 W. Jones, Nucleic Acids. Their Chemical Properties and Physiological conduct, London, 1920.

36 A. Kossel, Z. Physiol. Chem., 3 (1879) 284.
37 A. Kossel, Z. Physiol. Chem., 4 (1880) 290.
38 A. Kossel, Z. Physiol. Chem., 6 (1882) 422.
39 A. Kossel. Z. Physiol. Chem., 8 (1883—1884) 404.
40 A. Kossel, Z. Physiol. Chem., 10 (1886) 248.
41 A. Kossel, Z. Physiol. Chem., 12 (1888) 241.
42 C. Neubauer, Z. Physiol. Chem., 6 (1867) 33.
43 S. Schindler, Z. Physiol. Chem., 13 (1889) 432.
44 G. Bruhns, Z. Physiol. Chem., 14 (1890) 533.
45 A. Kossel, Ber. Dtsch. Chem. Ges., 18 (1885) 79.
46 N. Lubavin, Ber. Dtsch. Chem. Ges., 10 (1870) 2238.
47 G. Bunge, Z. Physiol. Chem., 9 (1885) 49.
48 A. Kossel, Arch. Anat. Physiol., (1891) 181.
49 L. Liebermann, Ber. Dtsch. Chem. Ges., 21 (1888) 598.
50 J. Pohl, Z. Physiol. Chem., 13 (1889) 292.
51 H. Malfatti, Z. Physiol. Chem., 16 (1892) 68.
52 A. Kossel, Arch. Anat. Physiol., (1893) 153.
53 E. Fischer, Ber. Dtsch. Chem. Ges., (1898) 2250.
54 E. Fischer, Untersuchungen in der Puringruppe, Berlin, 1907.
55 L. Medicus, Ann. Chem., 175 (1875) 243.
56 P.J. Pelletier and J.B. Caventou, Berzelius Jahresber., 4 (1825) 180.
57 R. Behrend and O. Roosen, Ann. Chem., 251 (1889) 235.
58 E. Fischer, Ber. Dtsch. Chem. Ges., 30 (1897) 2226.
59 A. Strecker, Ann. Chem., 108 (1858) 141.
60 A. Kossel, Z. Physiol. Chem., 10 (1886) 248.
61 R. Altmann, Arch. Anat. Physiol., (1889) 524.
62 A. Kossel and A. Neumann, Ber. Dtsch. Chem. Ges., 26 (1893) 2753.
63 A. Kossel and A. Neumann, Ber. Dtsch. Chem. Ges., 27 (1894) 2215.
64 A. Kossel, Arch. Anat. Physiol., (1891) 181.
65 A. Kossel and H. Steudel, Z. Physiol. Chem., 38 (1903) 49.
66 A. Ascoli, Z. Physiol. Chem., 31 (1900—1901) 156.
67 M. Stadthagen, Virchow's Arch., 109 (1887) 390.
68 G. Gumlich, Z. Physiol. Chem., 18 (1894) 508.
69 W. von Mach, Schmiedeberg's Arch., 24 (1888) 389.
70 J. Horbaczewski, Monatsh. Chem., 2 (1882) 796.
71 J. Weiss, Z. Physiol. Chem., 27 (1899) 216.
72 H. Sgeudel, Z. Physiol. Chem., 32 (1901) 285.
73 J. Horbaczewski, Monatsh. Chem., 10 (1889) 624.
74 J. Horbaczewski, Monatsh. Chem., 12 (1892) 221.
75 A. Kossel, Z. Physiol. Chem., 12 (1888) 241.
76 O. Minkowski, Schmiedeberg's Arch., 41 (1898) 375.
77 W. Spitzer, Arch. Ges. Physiol., 76 (1899) 192.
78 R. Burian, Z. Physiol. Chem., 43 (1904—1905) 497.
79 H. Wiener, Beitr. Chem. Physiol., 2 (1902) 42.
80 M. Freudweiler, Dtsch. Arch. Klin. Med., 69 (1900).
81 M. Nencki, I.P. Pavlov and J. Zaleski, Arch. Exp. Pathol. Pharmacol., 37 (1895) 26.
82 R. Behrend and O. Roosen, Ann. Chem., 251 (1889) 235.
83 J. Horbaczewski, Monatsh. Chem., 8 (1887) 201.

84 H. Meyer, Beiträge zur Kenntnis des Stoffwechsels im Organismus der Hühnen, Dissertation, Königsberg, 1877.
85 W. Pfeiffer, Ber. Dtsch. Chem. Ges., 10 (1907) 324.
86 R.B. Fischer, Biochem. J., 29 (1935) 2198.
87 F. Wöhler, Ann. Chem., 70 (1849) 229.
88 F.T. Frerichs and G.A. Städeler, Arch. Anat. Physiol., (1854) 393.
89 G. Meissner, Zeitschr. Rat. Med., 31 (1868) 144.
90 O. Folin, J. Biol. Chem., 69 (1924) 361.
91 F. Soetbeer and J. Ibrahim, Z. Physiol. Chem., 35 (1902) 1.
92 O. Minkowski, Arch. Exp. Pathol. Pharmacol., 41 (1898) 375.
93 T. Cohn, Z. Physiol. Chem., 25 (1898) 929.
94 E. Salkowski, Centr. Med. Wiss., 36 (1898) 929.
95 R. Poduschka, Arch. Exp. Pathol. Pharmacol., 44 (1900) 59.
96 L.B. Mendel and E.W. Brown, Am. J. Physiol., 3 (1899—1900) 261.
97 W. Wiechowski, Beitr. Chem. Physiol. Pathol., 11 (1907) 109.
98 W.C. Winternitz and W. Jones, Z. Physiol. Chem., 60 (1909) 180.
99 J.R. Miller and W. Jones, Z. Physiol. Chem., 61 (1909) 395.
100 A. Schittenhelm, Z. Physiol. Chem., 63 (1909) 248.
101 R. Behrend, Ann. Chem., 333 (1904) 144.
102 H. Biltz and F. Max, Ber. Dtsch Chem. Ges., 54 (1921).
103 H. Biltz and H. Schauder, J. Prakt. Chem., 106 (1923) 108.
104 H. Brünig, F. Einecke, F. Peters, R. Rabl and K. Viehl, Z. Physiol. Chem., 174 (1928) 94.
105 F. Battelli and L. Stern, Biochem. Z., 19 (1909) 219.
106 K. Felix, F. Scheel and W. Schuler, Z. Physiol. Chem., 180 (1928) 90.
107 W. Schuler and W. Reindel, Z. Physiol. Chem., 208 (1932) 248; 215 (1933) 258.
108 R. Fosse, P. De Graeve and P.E. Thomas, Compt. Rend., 197 (1933) 370.
109 S.J. Przylecki, Arch. Int. Physiol., 24 (1925) 237.
110 A. Nemec, Biochem. Z., 112 (1921) 286.
111 R. Fosse and A. Brunel, Compt. Rend., 188 (1929) 1067.
112 H.A. Krebs and H. Weil, Problèmes de Biologie et de Médecine. Volume Jubilaire Dédié au Prof. L. Stern, Moscow-Leningrad, 1935, p. 497.
113 A. Brunel, Le Métabolisme de l'Azote d'Origine Purique chez les Champignons, Thesis, Paris, 1936.
114 R. Fosse, Uréogenèse et Métabolisme de l'Azote Purique chez les Végétaux, Paris, 1939.
115 R. Echevin et A. Brunel, Compt. Rend., 204 (1937) 881; 205 (1937) 294.
116 M. Florkin, L'évolution du Métabolisme des Substances Azotées chez les Animaux (Actualités Biochimiques, No. 3), Paris, 1945.
117 R. Fosse, A. Brunnel and P. De Graeve, Paris, 1945. Compt. Rend. Sci., 190 (1930) 79.
118 A. Brunel, Bull. Soc. Chim. Biol., 19 (1937) 805.
119 E. Stransky, Biochem. Z., 266 (1933) 287.
120 A. Brunel, Bull. Soc. Chim. Biol., 19 (1937) 1027.
121 M. Florkin and G. Duchâteau, Arch. Int. Physiol., 52 (1942) 26.

122 M. Florkin, L'évolution Biochimique, Paris, 1944 (translation by S. Morgulis, Biochemical Evolution, New York, 1949).
123 S. Eppinger, Beitr. Chem. Physiol. Pathol., 6 (1905) 287, 492.
124 W. Pfeiffer, Beitr. Chem. Physiol. Pathol., 10 (1907) 324.
125 G. Ascoli, Arch. Ges. Physiol., 72 (1898) 340.
126 M. Ascoli and G. Izar, Z. Physiol. Chem., 58 (1909) 529.
127 C. Bezzola, L. Preti and G. Izar, Z. Physiol. Chem., 62 (1909) 229.
128 G. Izar, Z. Physiol. Chem., 64 (1910) 62; 65 (1910) 78; 73 (1911) 317.
129 L. Preti, Z. Physiol. Chem., 62 (1909) 354.
130 H.M. Spiers, Biochem. J., 9 (1915) 337.
131 H.O. Calvery, J. Biol. Chem., 73 (1927) 77.
132 G. Pupilli, Arch. Physiol., 26 (1930) 113.
133 G. Pupilli, Bioch. Terap. Sper., 14 (1930) 451.
134 A. Clementi, Arch. Sci. Biol., 14 (1930) 451.
135 D. Torrisi and F. Torrisi, Boll. Soc. Ital. Biol. Sper., 6 (1931) 262.
136 G. Russo and C. Cuscuna, Boll. Soc. Ital. Biol. Sper., 6, (1931) 250.
137 K. Kowaleski and S. Salaskin, Z. Physiol. Chem., 33 (1901) 210.
138 W. Schuler and W. Reindel, Z. Physiol. Chem., 221 (1939) 209.
139 W. Schuler and W. Reindel, Z. Physiol. Chem., 221 (1933) 221.
140 W. Schuler and W. Reindel. Z. Physiol. Chem., 221 (1933) 232.
141 T.H. Benzinger and H.A. Krebs, Klin. Wochschr., 12 (1933) 1206.
142 H.A. Krebs, Biochem. Educ., 1 (1973) 19.
143 N.L. Edson, H.A. Krebs and A. Model, Biochem. J., 30 (1936) 1380.
144 W. Reindel and W. Schuler, Z. Physiol. Chem., 248 (1937) 197.
145 A. Orström, M. Orström and H.A. Krebs, Biochem. J., 33 (1939) 990.
146 W.C. Rose, Physiol. Rev., 3 (1923) 544.
147 A.A. Christman, Physiol. Rev., 32 (1952) 303.
148 O. Folin, Am. J. Physiol., 13 (1905) 66.
149 A.E. Taylor and W.C. Rose, J. Biol. Chem., 18 (1914) 519.
150 G. Graham and E.P. Poulton, Quart. J. Med., 7 (1913—1914) 13.
151 L.B. Mendel and R.L. Stehle, J. Biol. Chem., 22 (1915) 215.
152 N. Umeda, Biochem. J., 9 (1915) 421.
153 H.B. Lewis, M.S. Dunn and E.A. Doisy. J. Biol. Chem., 36 (1918) 9.
154 A.A. Christman and E.C. Mosier, J. Biol. Chem., 83 (1929) 11.
155 H. Borsook and G.L. Keighley, Proc. Roy. Soc. London, Ser. B, 118 (1935) 488.
156 A.J. Quick, J. Biol. Chem., 98 (1932) 157.
157 H.V. Gibson and E.A. Doisy, J. Biol. Chem., 55 (1923) 605.
158 W.G. Lennox, J. Biol. Chem., 66 (1925) 521.
159 H. Ackroydt and F.G. Hopkins, Biochem. J., 10 (1916) 551.
160 H.B. Lewis and E.A. Doisy, J. Biol. Chem., 36 (1918) 1.
161 V.J. Harding and E.G. Young, J. Biol. Chem., 40 (1919) 227.
162 W.C. Rose, J. Biol. Chem., 48 (1921) 563.
163 W.C. Rose, and K.G. Cook, J. Biol. Chem., 6 (1925) 325.
164 P. György and S.J. Thannhauser, Z. Physiol. Chem., 180 (1929) 286.
165 W.A. Crandall and E.G. Young, Biochem. J., 32 (1938) 1133.
166 E. Abderhalden and H. Einbeck, Z. Physiol. Chem., 62 (1909) 322.
167 E. Abderhalden, H. Einbeck and J. Schmid, Z. Physiol. Chem., 68 (1910) 395.
168 E.G. Young, C.F. Conway and W.A. Crandall, Biochem. J., 32 (1938) 1133, 1138.
169 L.R. Cerecedo, J. Biol. Chem., 88 (1930) 695.

Chapter 50

Other Biosynthetic Aspects of Animal Chemistry

1. Introduction

Considering the approaches practised during the period covered in the present volume (before the introduction of isotopes, i.e. from the phlogistonic and pneumatic chemists to a period approaching World War II) with respect to biosynthesis, Chapters 47, 48 and 49 give us a picture of the methodological practices of the animal chemists. As reviewed in Chapter 46, organic chemists developed a large body of schemes based on the knowledge of the synthesis of organic constituents of plants in vitro. On the other hand, many physiologists were interested in plant nutrition and produced large amounts of printed reports on essential plant biosyntheses and photosynthesis, the value of which has become almost negligible (Chapter 45).

In the field of animal biochemistry, the organic chemists followed the practices of Liebig, Dumas, Pelouze, etc. in developing chemically based deductions concerning physiological processes of which it must be recognized they knew nothing. It was only after World War II that a fertile interplay between chemistry and biology was inaugurated. This development was favoured by undeniable successes of animal and plant chemistry. These achievements could not fail to impress rational minds.

Organic chemists were satisfied by considering the biosynthesis of urea as "explained" by the reproduction in vitro, of the reaction

$$\text{ammonium carbonate} \rightarrow \text{urea.}$$

Short-sighted as it may have been, this was a correct statement of a frame in which the discovery of the "ornithine cycle" was later situated through the brilliant work of Krebs and Henseleit.

Animal chemistry was first promoted predominantly by physiologists, clinicians and pathologists primarily concerned with human or animal diseases *.

While organic chemists were satisfied by reproducing in vitro a synthesis of uric acid from urea and another candidate (tartronic acid in the most popular theory), experimental researches in animal chemistry identified the "oxidative" pathway from purines to uric acid. As we have seen, the views of the pathologists (interested in the biosynthesis of uric acid since the discovery by Wollaston of uric acid in the tophous concretions of gouty patients) were obscured by their belief that the end-product of the puric metabolism of all mammals was uric acid, whereas in fact it is usually allantoin.

It took more work to show that, while in mammals the end-product of the metabolism of amino acids is urea, the end-product of the puric metabolism is generally allantoin (but uric acid in man and anthropoid apes). In birds the end-product of both amino acid and purine nitrogen metabolism is uric acid. In amphibia, urea is the end-product of both, whereas in lower invertebrates, ammonia is the end-product of both. The impact of animal experimentation was therefore clearly essential to define the beginning and the end of a biosynthetic pathway and to recognize its occurrence, and this aspect was outside the possibilities of the approach of organic chemists.

In spite of such methodological warnings and of the opposition of physiologists, and foremost among them Cl. Bernard, the physiological chemists who had been raised in the context of organic chemistry generally persisted until after World War II in their tradition. When the number of natural substances that were deciphered in their constitution and synthesis increased, these chemists adopted the methodology of testing, as to their excretion products by mammalian organisms, new compounds described by organic chemists. When the excretion product, as well as the precursor introduced, were linked by a process of synthesis in vitro, they considered the pathway as demonstrated. It was, as denounced by the physiologists, a short-sighted approach, but it must

* This aspect is illustrated by many interesting examples in Coley's book [1].

be granted that the organic chemists, however limited in their interpretations, remained unwarped by many irrelevant theories which misguided the physiologists, traditionally inclined towards finalist and organicist concepts.

Chemists were not involved in the vitalism-mechanism controversies which raged among physiologists and pathologists. The dislike for speculation was a feature of the chemist's approach (including, unfortunately, an aversion for the enzyme theory). While physiologists were prone to an excess of theorization, as exemplified for instance by Pflüger's theory of high energy proteins, based on a mere analogy, the sin of organic chemists, when dealing with aspects of physiological chemistry, consisted in an almost puritanical excess of caution, excluding all inductions which were not directly chemically based. This aversion for theories, and even for ideas, was, among other aspects, exemplified in the views of Comte, the founder of French positivism, who considered only factual evidence as valuable. If we take the reader ahead of the History as it unfolded, we cannot deny that the essential concept illuminating modern biochemistry, the notion of the intramacromolecular order, was formulated by organic chemists, and was not suspected by physiologists.

After World War II, organic chemists more generally (not universally) recognized that a mere comparison of chemical structures was not of a nature to inform about biosynthetic pathways. It was the task of the biochemists to determine these pathways by appropriate experiments on organisms. This change of philosophy was prepared by the impressive successes of animal chemistry, exemplified by the identification of the ornithine cycle by Krebs and Henseleit, or by the recognition, by Channon, of squalene as a precursor of cholesterol in vivo.

On the other hand, the tradition of organic chemists to reproduce in vitro the steps of a biosynthetic pathway became a firmer and firmer asset of the classical period of biosynthetic studies. Epistemologically, this derived from the chemical method of confirming a molecular structure by the process of synthesis, and it remained one of the most productive tenets of the conceptual system of biochemical methodology. Another obvious contribution of organic chemistry, in the context of the interplay of chemistry and biology, was the production, through the wide knowledge of natural substances, of the epistemological background which

Plate 169. Eugen Baumann.

formed the canvas on which the modern knowledge of biosynthetic pathways was embroidered, as will be documented in Part V of the present *History*.

This unravelling of biosynthetic pathways will depend on the progress of knowledge of chemistry and physics in the field of isotopes, on a progress of the chemistry of proteins and consequently of enzymes, as well as on the application of such biological concepts as that of auxotroph mutants of microorganisms. Such progress broadened the possibility of experimentation on organisms, as had been the wish of the physiologists, as well as the formulation and reproduction in vitro of the intermediary steps according to the tradition of organic chemists.

This fruitful interplay of biology and chemistry was the background of the impressive acceleration of knowledge of the biosynthetic pathways that took place after World War II.

2. Detoxication syntheses

In the course of the early period considered in the present volume, animal chemistry enhanced the knowledge of biosynthesis by a number of important contributions. One of these, nearest to the modern ideal of biochemists, was the accurate work on "detoxication syntheses" accomplished by Baumann as early as the 1870's, and in which animal experimentation and the reproduction of the biosynthetic products in vitro gave a preview, facilitated by the excretion in urine of the product (as had also happened with urea and uric acid), of the aims of biochemical research. Other excretion syntheses have also been completely or partially unravelled.

(a) Hippuric acid

Hippuric acid was discovered by Liebig in 1830. In his Faraday Lecture on Liebig, which he delivered before the Chemical Society of London in 1875, Hofman [2], tells that

"In 1830, Liebig repeated the experiments of Rouelle as well as the later ones of Fourcroy and Vauquelin which had indicated the presence of benzoic acid in the fluid secretions of horses. These experiments led him to the discovery of an acid containing nitrogen, to which he gave the name hippuric acid. He

References p. 302

remarks that it may be considered as a compound of benzoic acid with an organic body as yet unknown. The nature of this body was reserved for Dessaignes to establish, by showing that hippuric acid, when treated with acids, splits up into benzoic acid and glycocoll" (p. 84).

Here again, the many stimulations given to biochemical studies by the treatment of gouty patients appear in a further aspect. In 1841, a clinician, Ure [3], had the idea of administering to gouty subjects, an hour after a meal, benzoic acid. He then observed an excretion of hippuric acid which he considered as having replaced, by an easily soluble salt, sodium hippurate, the sparingly soluble sodium urate, responsible for the formation of the tophous concretions. Whatever judgement may be passed on such metabolic theory, the fact is that Ure was the first to observe the conversion, in vivo, of benzoic acid into hippuric acid. It is sometimes stated that Wöhler was the author of this discovery, but while he had conceived the idea that benzoic acid could be converted into hippuric acid in vivo, he confused in his experiments, as was stated by his student Keller [4], benzoic acid and hippuric acid. After Ure's publication, Wöhler suggested to Keller to experiment on himself, and Keller unequivocally confirmed the conversion. The presence of hippuric acid in urine was therefore explained by a biosynthetic process, reproducible in vitro and expressed by a chemical equation:

$$\text{benzoic acid} + \text{glycine} \rightarrow \text{hippuric acid.}$$

(b) Conjugated sulphates

As Lieben [5] relates (p. 607), it was known since the middle of the 18th century that both normal and pathological urines sometimes give, on standing, a bluish sediment. This sediment had been named "uroxanthine" by the Viennese clinician J.F. Heller and had been identified by Dumas as indigo blue. The chromogen of this substance became known as "indigo-forming substance". A similarity had been suggested by E. Schunck between some plant glycosides (for instance of *Isatis tinctoria* and other indigo-forming plants) and the "indigo-forming substance" of urine, and it was common thereafter to find in the literature the expressions "urinary indican" and "plant indican" as synonymous with "indigo-forming substances".

In Hoppe-Seyler's laboratory, in Strassburg, Baumann [6] found that the "indigo-forming substances" of urine were decomposed by glacial acetic acid and by hydrochloric acid to give sulphuric acid. He concluded that these substances were "paired sulphates" with a carbon-containing substance of unknown nature.

Baumann, the son of an apothecary, had received a thorough chemical education at the Polytechnikum of Stuttgart. Though he did not study medicine, he was deeply interested in the problems of pathology. Completing his studies in Tübingen, Baumann was examined in toxicology by Hoppe-Seyler who was so impressed with his knowledge that he asked him to become his assistant. Baumann therefore became one of those who worked in the former kitchen of the old Tübingen Castle in which Hoppe-Seyler's department was housed. In 1872, Hoppe-Seyler became professor of physiological chemistry in Strassburg, a city annexed by the Germans as a consequence of their victory over the French in the war of 1870. Baumann followed his master to Strassburg, and he accomplished there his work on conjugated sulphates. In 1877, Baumann became chief of the chemical section of the Physiological Institute of Berlin, whose director was Du Bois-Reymond. From there, he moved to Freiburg as professor of chemistry at the Medical Faculty. He was not yet 50 years old when he died, in 1896, of a heart attack [7].

Baumann [8] recognized the "paired sulphates" as derivatives of phenol, catechol and the "indigo-forming substances". Potassium phenylsulphate was isolated by Baumann in the crystallized form from the urine of a patient treated with carbolic acid. He also observed an increase of conjugated sulphates in the urine of a dog that had received 2 g of catechol.

Williams [9] has pointed to the fact that the existence of two forms of sulphate in urine had already been noticed by earlier chemists and was for instance mentioned in the widely read treatise of Henry [10]. The presence of conjugated sulphates in urine was thus explained by a chemical equation

$$\text{phenol} + \text{sulphuric acid} \rightarrow \text{paired sulphates.}$$

(c) Glucuronoconjugation

In line with the practice of administering organic compounds to animals and gathering excretion products, it was observed that

References p. 302

feeding, for instance, chloral hydrate, nitrobenzene, morphine etc. resulted in the appearance, in the urine, of a laevorotation and of a reducing power. Glucose was ruled out by the observation that fermentation did not abolish these new properties of urine. After administering chloral hydrate to a human subject, von Mering and Musculus [11] isolated from his urine an acidic, reducing, laevorotatory compound, which they called "urochloralic acid".

In a study on the metabolism of *o*-nitrotoluene, Jaffé [12] observed among other products, that the main one was a laevorotatory acidic compound, which he called "uronitrotoluol acid". In 1879, Schmiedeberg and Meyer [13] isolated from the urine of dogs, after an ingestion of camphor, a compound from which they separated a substance they considered as corresponding to the hypothetical acid of Jaffé and partly to the "urochloralic acid" of Mering and Musculus. The new substance, isolated in conjugation with camphor, was called "glycuronic acid" by Schmiedeberg and Meyer, who proposed for it the structure

$$(CHOH)_4 \begin{cases} CHO \\ \\ COOH \end{cases}$$

They considered it as a catabolic product derived from glucose. A hypothetical pathway, resulting from guesses derived from chemical knowledge, was proposed by Sundwik [14] and by Fischer and Piloty [15] who proposed that the substance was first combined with glucose, forming a glucoside, and subsequently oxidized to glucuronide. As it was known that glucosides are not oxidized to glucuronides in animals, the key to the formation of glucuronides was only found much later.

(d) Mercapturic acid

In 1879, Baumann and Preuse [16] on the one hand and Jaffé [17] on the other hand, observed that after feeding bromobenzene or chlorbenzene, respectively, to dogs their urine contained mercapturic acids, i.e. derivatives of *N*-acetyl-L-cysteine

$$\begin{array}{l} R-S-CH_2-\underset{|}{C}H-COOH \\ \qquad\qquad\quad\ \ NH-COCH_3 \end{array}$$

(R = aryl or alkyl group)

Only in the 1950's was the biochemical mechanism unravelled.

(e) The "detoxication concept"

Although he noted, in one of his papers [6] of 1876, that the phenols are toxic and that phenyl-sulphates are not, Baumann continued to classify such biosyntheses as those of urea, uric acid, hippuric acid, conjugated sulphates, etc. as "excretion syntheses". The concept of "detoxication", well in line with the finalist attitudes of physiologists, was not accepted by Baumann, nor by his fellow physiological chemists. As noted by Williams [9], one of the first presentations of the concept of detoxication (Entgiftung) is found in the *Lehrbuch* [18] of Neumeister, with reference to glucuronoconjugation.

3. "Excretory synthesis" and "intermediary synthesis"

As we have seen in Chapter 48, Nencki (1872) had suggested that the excretion syntheses, as well as other syntheses, were the result of an abstraction of a molecule of water from the two substances that are combined. As stated by Fruton [19] (p. 412), the concept of the metabolic importance of the loss or gain of the elements of water was formulated by Baeyer [20] in 1870, with respect to plant physiology. The recognition of the different excretion syntheses enumerated above pointed to the formation of an anhydride as a biosynthetic mechanism.

In 1878, Baumann [21] delivered an important lecture in which, reviewing the knowledge acquired on what he called "excretion syntheses", he envisaged the possibility of invoking the anhydride formation to explain what he called "intermediary syntheses" such as for instance the formation of glycogen from glucose or the formation of proteins from "peptones". Baumann referred to the theory of anhydrides as a possible explanation although, with his usual caution, he did not reach conclusion on the matter. Fruton [22] as well as Spaude [23] pointed out that Baumann carefully avoided speculation.

The extension of the anhydride theory to the formation of such compounds as glycogen was in the forefront of metabolic theories for several decades. While one theory considered that glycogen formation results in the loss of water from dietary glucose, there was

References p. 302

a theory, originating in Bernard's classical experiments (see Chapter 10) according to which glycogen could derive from compounds other than carbohydrates. This view was first repudiated by Pflüger [24] in a long classical paper on glycogen, almost as long as a book. Studies on experimental diabetes convinced Pflüger of the possibility of a formation of glycogen from proteins, and he adhered to this concept in a short note in 1910 [25].

Among their experiments on diabetic dogs, Stiles and Lusk [26] gave to a diabetic dog a mixture of amino acids prepared by digestion of meat with pancreatic juice. They found an increase in the elimination of sugar corresponding to about 40 per cent of the amino acids administered. Embden and Salmon [27], fed glycine and alanine to depancreatized dogs and obtained a considerable increase in the elimination of glucose. Experiments were also performed on dogs presenting "phloridzin diabetes". This artificial diabetes is produced by the plant glycoside phloridzin, lowering the threshold of glucose excretion by the kidney. It had been introduced by von Mering as a method for the identification of precursors of glucose. In a phloridzinized dog fed on meat, Knopf [28] obtained a decided increase in the elimination of glucose after adding 50 g of asparagin to the diet. Ringer and Lusk [29], working on phloridzinized dogs observed that glycocoll and alanine may be completely converted into glucose. Mandel and Lusk [30] showed that lactic acid may be completely converted into glucose.

Ringer, from the Department of Physiological Chemistry of the University of Pennsylvania, started in 1912, the publication of a series of scientific articles in the Journal of Biological Chemistry under the general title "The chemistry of gluconeogenesis". That pyruvic acid was glucogenic was shown by Ringer [31] and by Dakin and Janney [32].

It became agreed that glucose, which could be derived from amino acids or other substances, was the precursor of glycogen. The suggestion that the glycogenic activity of proteins may result from their alleged association with carbohydrates, was refuted by experiments with highly purified proteins free of carbohydrates. That gluconeogenesis not only occurs in liver but also in kidney was demonstrated by Benoy and Elliott [33] in 1937. Renal gluconeogenesis from lactate, pyruvate, amino acids and the intermediates of the tricarboxylic acid cycle was first observed by these authors on tissue slices. It was confirmed by Weil-Malherbe [34] during the following year.

It was thus accepted that the liver and the kidney cortex are the main sites of gluconeogenesis. As mentioned in Chapter 20, Meyerhof had postulated a conversion of lactate into carbohydrate in muscle (Meyerhof cycle), but this theory was not confirmed. It was later interpreted, as we shall see, by a relative or complete lack of relevant enzymes in muscle tissues. As Cori and Cori [35] showed in 1929 in their classical work, the resynthesis of glucose from lactate involves the "Cori cycle". The lactate is transferred by circulation to liver and to kidney cortex, where glucose is formed by the very complex pathway of gluconeogenesis which was only interpreted after World War II. The notion of the polymerization of glucose into glycogen became, in the first decades of this century, one of the aspects of the current concept, discussed in Chapter 51 hereafter, of biosynthesis by reversibility of hydrolytic enzymic action.

4. The biosynthesis of fats

We have retraced, in Chapter 42, the controversies that led to the recognition of a biosynthesis of fats by animals. In 1862, H.M. Edwards [36] nevertheless still maintained (p. 550) that

> "The major part of the fat which accumulates in animal economy preexists in food and is introduced in the organism by way of the digestive tract". (Author's translation.)

As stated in Chapter 42, chemical equations had been imagined and experiments had been performed in vitro with the view of explaining the formation of fats in the animal body. Boussingault demonstrated the ability of pigs to accomplish the biosynthesis of fats.

The first experimental proof of a conversion of carbohydrate to fat was provided by Lawes and Gilbert [37] in 1866 and by Meissl and Strohmer [38] in 1883. The authors accomplished the determination of balance sheets in pigs and concluded that storage of fat derived from the carbohydrates of the diet indeed occurred. Balance experiments of similar nature confirmed the formation of fats from carbohydrates in geese (Lehmann and Voit [39]) and

References p. 302

in dogs (Rubner [40,41]). To these results of the application of an indirect method were added those of another indirect method, the elevation of the respiratory quotient, CO_2/O_2, after the ingestion of a diet rich in carbohydrates. (Literature in Deuel [42], pp. 539—540.)

Though no direct study of the pathway was possible before the introduction of deuterium as a marking device for following the fate of molecules, the knowledge that the chief fatty acids synthesized were the C-16 and C-18 acids, units of 2 carbons were logically considered as building stones. The intermediate between carbohydrates and fats was suggested by Magnus-Levy [43] as being acetaldehyde. He proposed, quite hypothetically and without a reproduction of these reactions in vitro, that two molecules of acetaldehyde condense to form hydroxybutyral which is reduced to butyraldehyde. Successive condensations of acetaldehyde with butyraldehyde, followed by the reduction of β-hydroxyl groups, lead to the formation of long chains and an oxidation of the terminal aldehyde group determined the final length. Such hypothetical schemes were reinforced by the knowledge of the presence of the aldehydes of the higher fatty acids in normal tissues (Feulgen, Imhäuser and Behrens [44]).

That proteins can be converted to fat was indicated by balance experiments accomplished by Pettenkofer and Voit [45], Lawes and Gilbert [37] and Meissl and Strohmer [38]. In their balance experiments, Pettenkofer and Voit [45] had calculated the carbon content of proteins on the basis of the nitrogen content, using the factor 3.68 for the C/N ratio. Rubner [46] determining this factor on meat completely extracted with ether, concluded that the value for the C/N ratio was 3.28. Recalculating the results of Pettenkofer and Voit on this basis, Pflüger [47] concluded that the carbon retained was insignificant and he denied the transformation of protein into fat.

However, the experiments performed by Cremer [48] on cats provided an inescapable demonstration of this conversion. This principle was substantiated by many data (literature in Deuel [42], vol. II, pp. 549—553) proving the conversion of proteins into carbohydrates.

5. The biosynthesis of cholesterol

(a) Discovery

In 1789, Fourcroy [49] published the results of a study of some gall-stones and urinary stones. F. Poulletier de la Salle had, according to Fourcroy, observed in 1770 that gall-stones were almost completely soluble in alcohol, but Fourcroy was the first to obtain a crystallized material from gall-stones. He dissolved them in warm alcohol, filtered the solution, and on cooling the filtrate obtained white crystals. This material was evidently still very impure because crystals melted below boiling water temperature, whereas the melting point of pure cholesterol is known to correspond to 119°C. The substance isolated from gall-stones was insoluble in hot or cold water, but dissolved in alkalis, forming soapy solutions. As stated above, it was soluble in hot alcohol.

Fourcroy had the opportunity of comparing these properties with those of a material similar to spermaceti which he found in exhumed bodies when he was led to explore the corpses dug up from the Parisian "Cemetery of the Innocents". This huge graveyard, mentioned by Rabelais, was, since 1186 enclosed by galleries lined with charnel-houses in which Vesalius had found bony specimens for his anatomical studies. In 1782 this graveyard was transformed into a market place, the location of which corresponds to the present "Square of the Innocents" in the quartier des Halles, described by Zola in his famous novel *Le Ventre de Paris*.

Together with a well-known medical authority, Thouret (see Florkin [50], pp. 171–173), who later became Director of the Medical School of Paris, Fourcroy [51] was commissioned to report on the corpses found in the cemetery, on which he performed a series of chemical investigations during the years 1786 and 1787. The bodies, buried in individual graves, were putrefied after a few years in damp earth, and mummified in dry earth. But most bodies were buried in communal graves. In those, up to 1500, the coffins had been piled side by side. In these corpses, Fourcroy was surprised to find that, if the contents of abdomen and thorax were completley putrefied, muscle, fat and skin were converted into a white greasy material which he called "gras de cadavre" (fat of corpses). Fourcroy interviewed the grave diggers,

References p. 302

who were well aware of this conversion and who told him that it was generally accomplished within five years. They also told him that fat was the first tissue to be converted and muscle the last. "Gras de cadavre" was moist and soft at first and later became dry and friable. Fourcroy [52] immediately drew a comparison with an observation he had made in 1785 on a piece of human liver which had remained exposed to the air for ten years. When, according to the usual procedure of the time, Fourcroy dry distilled this piece of putrefied liver, he obtained an oil that solidified on cooling and he observed marked effects of solvents on the solidified material. While only sparingly water-soluble, in the form of an ammoniacal soapy solution, this oil was completely dissolved in caustic potash with the liberation of ammonia. On the other hand, the piece of putrefied liver was almost completely soluble in alcohol. Adding water to this alcoholic solution, Fourcroy obtained the precipitation of an oily substance. The same oily substance obtained by heating the putrefied liver at 68°C crystallized on cooling. Fourcroy pointed to the similarity with spermaceti though he noted the latter's lower solubility in alcohol. Fourcroy, when he separated an oily solid from the "fat of corpses", compared it with the substance he had obtained from the putrefied liver, and with spermaceti. Noting that the substance was not found in living matter, he pointed to its resemblance to gall-stones. Before he performed his study on corpses, he had already [49] called attention to the similarities of the products obtained from the putrefied liver on the one hand and on the other hand of spermaceti and gall-stones. He suggested that the function of the mammalian liver was to excrete such waxy material which, when in excess, was at the origin of gall-stone formation.

In 1790, Fourcroy [53], from the data on solubility and melting point, classified spermaceti, "fat of corpses" and the white substance in gall-stones in the same category to which he gave, in 1802, in a general review on animal calculi and concretions, the name of "adipocire" [54]. In Chapter 5, we retraced the pioneer contribution of Chevreul on the characterization and analysis of fats. Chevreul published a series of papers on fats, the fifth of which [55] (in 1815) was concerned with the "adipocire" of Fourcroy. This paper is epitomized by Costa [56] as follows:

"Chevreul next investigated the different bodies which chemists referred to as varieties of "adipocire": the fat of corpses, spermaceti, and the crystalline substance present in human biliary calculi. Chevreul found that the crystalline substance present in human biliary calculi was not a fat. It did not saponify and, since other bodies could not be separated from it, it was a true immediate principle. On the other hand, spermaceti did saponify, although Chevreul found that the sweet principle was not among the saponification products. Finally, the fat of corpses was only a mixture of fats. Hence, Chevreul had found that these three bodies, each considered by Fourcroy to be varieties of the same principle, were really three distinct materials" (pp. 49—50).

These materials were given names in the 6th memoire [57] of Chevreul's series in which he devised a system of nomenclature. The immediate principle, unsaponifiable, crystallized from gall-stones, he called "cholesterine" meaning "solid bile" (from the Greek *khole*, bile, and *steros*, solid). Spermaceti he called "cétine", the acid from saponified cétine being called "cétic acid". In a paper published in 1817, Chevreul [58] described the purification of the spermaceti fat, cétine, its saponification into cétic acid and the recognition of this cétic acid as a mixture of margaric and oleic acids. The sweet principle of spermaceti fat was crystallized. In a later publication this neutral substance was called "ethal".

In his book of 1823, Chevreul [59] retraces the steps of the abrogation of Fourcroy's concept of "adipocire":

"Poulletier de la Salle was the first to isolate cholesterin by treating human gall stones by boiling alcohol. Fourcroy has compared this substance with a fatty material he obtained in 1785 from a human liver which had spontaneously putrefied in the presence of air, and with another fatty material obtained by him in 1786 from buried corpses. He has considered these three materials, as well as cetin as belonging to a same category of compounds to which he gave the name of "adipocire".

"In the paper presented by me to the Institut on September 19, 1814 I showed that cholesterin essentially differs from cetin and from the fat of corpses. I characterized these three substances by their very different behaviour in the presence of potash: see Book IV, Chapter IV" (pp. 159—160). (Translation by author.)

Chevreul [60] returned to this history in his book of 1870, dedicated to his master Vauquelin. Referring to his approach through the analysis of saponified fats and through the isolation of fatty acids before attempting the isolation of neutral fats, he recognized that the discovery of the liberation of glycerol from all those neutral fats and of a specific fatty acid for each neutral fat greatly simplified his work.

References p. 302

"Once this notion was established, it was easy to recognize the error of Fourcroy when he mingled under the common denomination "adipocire", the fat of corpses of the Innocent Cemetery, spermaceti and the crystallisable component of gall stones. On the basis of my previous studies, it became easy to distinguish these three materials one from another. Spermaceti, which I call "cetin" and the crystallisable material of gall stones, which I call "cholesterin" are neutral. Cetin melts at 48°C and gives, when saponified, margaric and oleic acids, and in addition ethal. Cholesterin, melting at 137°C, is not saponifiable. Finally the fat of corpses is acid and essentially composed of margaric and oleic acids" (pp. 184—185). (Translation by author.)

Later, cholesterin turned out to be recognized as a general constituent of animal tissues. A similar material was obtained from plants and called phytosterin. Later still, these compounds were recognized as belonging to the category of "sterols".

(b) Configuration

In 1888, Reinitzer [61] published a correct empirical formula of cholesterol. Cholesterol was converted into cholestenone by Diels and Abderhalden [62], who showed that the alcoholic group is secondary. That the molecule of cholesterol contains a double bond was shown by Wislicenius and Moldenhauer [63] in 1868.

The elucidation of the configuration of the cholesterol molecule was due to painstaking work, the stages of which have been retraced by Windaus [64]. Parallel studies had been performed by Wieland on the chemically related bile acids (series of papers entitled: Untersuchungen über die Konstitution der Gallensäuren, in Z. Physiol. Chem.). A tentative formula proposed by Windaus in 1919 was replaced in 1932 at the suggestion of Rosenheim and King [65] by a new formula. It was based on X-ray studies by Bernal [66] and on the discovery by Diels and Gädke [67] of the formation of chrysene by catalytic dehydrogenations. The final formulation was agreed upon by Windaus [64] and by Wieland and Dane [68]. This formula contains 8 asymmetric centres.

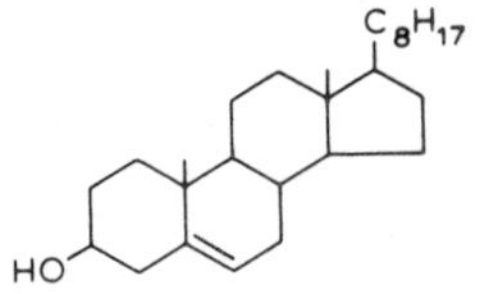

C_8H_{17}

HO

(c) Biosynthesis

That cholesterol is biosynthesized in rats fed with a cholesterol-free diet was shown by Dezani [69] and by Dezani and Cantoretti [70]. This biosynthesis was also observed in infants (Gamble and Blackfan [71]), human adults (Gardner and Fox [72]), chicks (Dam [73]) and laying hens (Schoenmeier [74]) but not in insects (Hobson [75]).

(d) Squalene recognized as a precursor of cholesterol in vivo

Squalene is a component of fish-liver oils, which have been the subject of much interest as a source of vitamin A. Squalene was discovered in 1906 by Tsujimoto [76]. As shown by Heilbron, Kamm and Owens [77], squalene is a long-chain hydrocarbon ($C_{30}H_{50}$) which they considered as a possible precursor of cholesterol. In fact, Channon [78], feeding squalene to rats, observed an increase in the cholesterol of their tissues. This discovery opened the way to future progress in the field and must be considered as a milestone.

(e) Biosynthesis of isoprenoids

In 1931, squalene was synthesized by Karrer and Helfenstein [79] who showed it to be a triterpene with six isoprene units. This knowledge situated the biosynthesis of squalene, precursor of cholesterol in vivo, into the frame of the biosynthesis of terpenes (a name derived from turpentine), a group of fragrant steam-volatile components of the essential oils of plants. H.H. Hlasiwetz who, when he was professor in Innsbruck, acquired fame through his studies which led to the isolation of amino acids from proteins, moved in 1867 to the Technische Hochschule of Vienna. There he obtained, in 1868, with heat, a decomposition of turpentine oil and of other plant oils with production of what he identified as the unsaturated hydrocarbon isoprene (C_5H_8) related to isopentane. This compound had been isolated by Faraday in 1826 among the products resulting from the dry distillation of rubber. Since the work of Hlasiwetz, terpenes have become defined, rather than

References p. 302

Plate 170. Paul Karrer.

with a reference to essential oils, by a more general criterion, their architectural and chemical relation to isoprene, C_5H_8. It must be noted that Berthollet had already, in 1860, commented on the periodicity of the architecture of some terpene molecules.

What became known as the "isoprene rule" (that some natural compounds can be divided into isoprenoid units) was applied with great success by Wallach in the interpretation of the architecture (Literature on the history of the "isoprene rule", in Ruzicka [80].) The application of the isoprene rule gave rise to a number of guesses about the biosynthesis of isoprenoids but, as we shall see, it was only after World War II that the subject made rapid progress.

(f) Hypothetical mechanism for the conversion of squalene into cholesterol

After the general acceptance of the Rosenheim-King formula of cholesterol, Robinson [81] proposed a hypothetical folding mechanism for the conversion of the linear squalene molecule to the tetracyclic sterol nucleus. It was only after the introduction of isotopes into biochemical methodology that the implementation of another mechanism was recognized, as we shall see.

6. The biosynthesis of porphyrins

(a) Discovery

In 1867, Thudichum [82] prepared, by the action of sulphuric acid on haemoglobin, the first porphyrin known, and called it cruentine. He described the absorption spectrum correctly. In 1871, Hoppe-Seyler [83] rediscovered the cruentine of Thudichum and called the purple pigment haematoporphyrin (from the Greek *porphyros* = purple).

In 1901, Nencki and Zaleski [84], treating haemin with HBr, obtained pure haematoporphyrin hydrochloride. By treating haemin with HI they obtained mesoporphyrin.

(b) Dualism of the porphyrins

In 1915, Hans Fischer was working, as a young chemist, in the clinical laboratory of Fr. von Müller in Munich. A patient named Petry, famous as the classical case of the disease named congenital porphyria excreted large amounts of undefined pigments. The disease, also called acute porphyria, had been described by Günther in 1912. The symptoms consisted in periodic attacks of abdominal pains and in mental troubles (see Dobriner and Rhoads [85]). Fischer isolated coproporphyrin from the urine of Petry. The excreted porphyrins, either in normal subjects or in those suffering of congenital porphyria were found to be different from protoporphyrin, the porphyrin of haemoglobin. ("Dualism of the porphyrins", see Fischer and Orth [86].) A survey of the finding of porphyrins in disease has been prepared by Dobriner and Rhoads [85].

(c) Synthesis

The synthesis of protoporphyrin and of haem (its iron complex) was accomplished by Fischer and Zeile [87] in 1929. This proved the tetrapyrrolic nature of the porphin nucleus, as well as the structure of many of its derivatives, the porphyrins. (See Fischer and Orth [86] and also Chapter I, by J.E. Falk, of Volume 9 of this Treatise.)

(d) Biosynthesis

The old literature on porphyrins is collected in the review of Dobriner and Rhoads [85].

Evidence was available that animals could biosynthesize haemoglobin without being fed with porphyrins. Chemical knowledge suggested glutamic acid as a possible precursor, because it can be converted into a derivative of proline. Proline and tryptophan were also suggested, because they contained rings similar to that of pyrrole. Fischer had suggested that acetoacetic acid might be a precursor, but by the time of the writing of the review of Dobriner and Rhoads [85] only speculative schemes had been proposed. (See also Rimington [88]; Turner [89].)

7. Creatine phosphate, creatine, creatinine

(a) Introduction

In Chapter 23, we retraced the discovery of creatine phosphate by Fiske and Subbarow, who isolated it from muscle in the form of its crystalline calcium salt. As stated in that Chapter, creatine phosphate is the "phosphagen" of vertebrate muscles and its role in the energetics of muscle contraction has been clearly defined.

The identification of creatine phosphate and of its role also gave the solution of an old query: the role of creatine, which had, since its discovery in 1832, remained mysterious. The biosynthesis of creatine had been the subject of a number of researches of animal chemists. The early studies concerning this biosynthesis may appear to the reader as obsolete material which does not deserve the attention of the historian. Admittedly these efforts do not belong to the category of errors that were ultimately revealed as fruitful. But the tentative interpretations of the biosynthesis of a compound which at the time appeared as an exceptionally large molecule, illustrates the futility of chemical schemes, even well founded, in the absence of a knowledge and identification of the "surface" at which they can properly be situated. As long as the physiology of creatine and creatinine was not properly situated in the frame of a vertebrate organism, tentative inductive applications of chemical knowledge remained futile.

When enough knowledge was acquired by animal chemists on the one hand and organic chemists on the other, it became possible, as will be shown in Part V, to form hypothetico-deductive views which could be tested by the application of new methods of enquiry, such as the isotopic method. The interpretation of such tests depended on a correct interpretation of the metabolic interrelations of creatine phosphate, creatine and creatinine, in relation to the knowledge of their chemical properties.

(b) The discovery of creatine phosphate

In 1922, A.P. Briggs [90] described a modification of the Bell-Doisy method for phosphate determination. In Brigg's procedure,

References p. 302

advantage is taken of the colour resulting from the reduction of phosphomolybdate. He allowed the colour to develop in acid solution during half an hour, whereas in the original Bell-Doisy method, this took place in slightly alkaline solution. Although the increase of colour during the half hour of its development corresponds to about 5 per cent, Eggleton and Eggleton, applying Briggs's method to extracts of frog's resting muscle prepared according to the procedures inaugurated by Meyerhof in 1926 (see Chapter 19), observed that the increase of colour developing in the course of the application of the method corresponded to several hundred per cent. The labile compound liberating phosphate in acid medium they called "phosphagen" in a paper which appeared in 1927 under the title: "The inorganic phosphate and a labile form of organic phosphate in the gastrocnemius of the frog" [91].

Independently, Fiske and Subbarow also reported, in 1927, on a labile phosphorus compound of muscle. Their background was also the application of Briggs's method. In 1925 they [92] described a rapid form of this method, requiring only four minutes for the development of colour. When, in 1927, they [93] applied their method to muscle extract, they observed that the colour was only fully developed after half an hour. From this they concluded that a labile phosphorus compound was present. In this paper of 1927 they recorded that the liberation of the labile phosphate was always accompanied by an equivalent amount of creatine.

In papers of 1928 and 1929, in the titles of which appears the term "phosphocreatine", Fiske and Subbarow [94,95] described the isolation of the substance from muscle as the crystalline calcium salt. Indeed, before that time, data had been suggesting an interrelationship between muscle creatine and other components of muscle tissue. Urano [96], in 1906, had already pointed out that in vitro, creatine and phosphate diffused from muscle at the same rate. This suggested the existence of a creatine-containing complex. But the first experimental evidence that, in vivo, muscle creatine is not free creatine is to be found in the work of Folin and Denis [97]. These authors showed, in 1914, that creatine injected into the circulation accumulates in muscles against an apparent concentration gradient, which pointed to the probable combination of intramuscular creatine.

In 1922, Embden and Adler [98] observed that frog muscle,

when bathed by Ringer solution, lost phosphoric acid very slowly when resting, but more rapidly when fatigued. Later, Tiegs [99] found the same effect for creatine. He interpreted this as due to a chemical change in creatine (which he considered as the presence of a nondialysable tautomer of creatine in resting muscle). The interpretation of Embden and Adler was different: they considered fatigue as accompanied by an increase of permeability of the membrane surrounding the muscle fibre. The idea that frog muscle is relatively impermeable to creatine phosphate resulted from the experiments of Embden and Adler and of Tiegs taken in conjunction with the observation of Stella [100] that, for a given concentration gradient, inorganic phosphate neither enters nor escapes more rapidly from muscle in the fatigued state.

Tiegs was right, as noted by Fiske and Subbarow

> "in ascribing the difference between resting and fatigued muscle to a chemical change involving creatine, even if he was entirely mistaken in the premises which led him to make this suggestion and in the nature of the change" (p. 678).

As was stated in Chapter 23, Lehmann [101] showed that the Parnas reaction

Phosphopyruvic acid + creatine ⇌ creatinephosphoric acid

+ pyruvic acid

ensures the formation of creatine phosphate in muscle. This was also demonstrated for other tissues by Torres [102] in Meyerhof's laboratory. She confirmed the process for testicle, heart, central nervous system, etc.

(c) The chemistry of creatine and creatinine

Creatine was isolated by Chevreul from meat extract in 1832, and this result was published in 1835 [103]. Chevreul obtained it in the form of transparent rectangular prisms. During the following decade several authors tried in vain to confirm Chevreul's finding. It was Liebig who, in 1847 [104,105], obtained creatine from the muscles of the mammals, birds and fishes he examined. The conclusion of Liebig, that creatine was a normal constituent of muscle in higher animals, was confirmed for all classes of vertebrates

References p. 302

(whale, Price [106]; fishes and birds, Gregory [107]; man, Schlossberger [108]; amphibia, Grohé [109]; reptiles, Schlossberger [110]).

In his first publication on creatine, in 1847, Liebig established its empirical formula. In the same paper he showed that by heating creatine with mineral acids, a new compound, which he called creatinine, was obtained by the loss of two molecules of water. On boiling with baryta, he obtained urea and a basic substance having the composition $C_4H_7ON_3$. To this new compound he gave the name sarcosine (Liebig [105]).

In 1854, Dessaignes [111], heating creatine with mercuric oxide, obtained a strong base, $C_2H_7N_3$, which he called "methyluramine". This "methyluramine" was recognized by Strecker [112] as methylguanidine. Strecker [113] had, in 1861, discovered among the products of the oxidation of guanine by hydrochloric acid and potassium chlorate a new base which he called guanidine (CH_3N_3). Strecker proposed that creatine was a combination of cyanamide $CN \cdot NH_2$ and a methylglycocoll, sarcosine. Bringing together aqueous solutions of cyanamide and of glycine and a few drops of ammonia, he obtained an addition product which he called glycocyamine.

$$\underset{\text{cyanamide}}{CN \cdot NH_2} + \underset{\text{glycine}}{C_2H_5O_2N} = \underset{\text{glycocyamine}}{C_3H_7O_2N_3}$$

On the basis of Strecker's synthesis of glycocyamine, Volhard [114], in 1868, accomplished the synthesis of creatine by reacting cyanamide and sarcosine

$$\underset{\text{sarcosine}}{C_3H_7NO_2} + \underset{\text{cyanamide}}{CN \cdot NH_2} = \underset{\text{creatine}}{C_4H_9O_2N_3}$$

After a period of confusion (see Hunter [115]) these observations were correctly interpreted by Strecker [112], followed by Erlenmeyer [116] who represented creatine as methylguanidineacetic acid, as we do today, and creatinine as its lactam (internal anhydride)

```
H3C—N—CH2                 H3C—N—CH2
    |   |                     |    |
HN=C   COOH               HN=C     |
    |                         |    |
   H2N                        HN—CO
```

methylguanidine-acetic acid (creatine) — glycine-methylguanidine (creatinine)

That creatine is methylguanidineacetic acid was confirmed by Horbaczewski [117], who, heating guanidine carbonate with sarcosine at 140—160°C, acidified with HCl the aqueous solution of the product, evaporated it to a syrup and obtained creatinine. This confirmed that creatine and glycocyamine were not substituted ureas but substituted guanidines.

$$\mathrm{NH{:}C}\begin{matrix}\diagup\ \mathrm{NH_2}\\ \diagdown\ \mathrm{NH_2}\end{matrix} + \begin{matrix}\mathrm{CH_3{\cdot}NH{\cdot}CH_2}\\ |\\ \mathrm{HO{-}CO}\end{matrix} = \mathrm{NH{:}C}\begin{matrix}\diagup\ \mathrm{N(CH_3){\cdot}CH_2}\\ \qquad\quad |\\ \diagdown\ \mathrm{NH{-\!\!-}CO}\end{matrix} + \mathrm{NH_3} + \mathrm{H_2O}$$

Horbaczewski's synthesis was confirmed, a decade later, by Paulmann [118].

(d) Creatine and creatinine in animal chemistry

Normal adult men do not excrete creatine in appreciate amounts when they receive a diet free from preformed creatine or creatinine, or a diet containing creatine unless in large amounts. They do not excrete unaltered the creatine circulating in their blood. Yet creatine is always found in the urine of normal children [119, 120]. Whereas children excrete creatine and creatinine, adult males excrete only creatinine. The younger the normal child, the less its ability to retain creatine given per os. This tolerance is reduced to a minimum in patients suffering from progressive muscular distrophy. Normal adult women show an intermittent creatinuria (Krause [121]). A large amount of data has been collected about the constant creatinuria of children and the occasional creatinuria of women. References to this literature can be found in the book by Hunter [115], who has also enumerated the various interpretations proposed.

We know now that creatine, produced in the liver (but not in muscle), circulates in the blood and is captured by muscle for the formation of one of its important constituents, creatine phosphate. The muscle system of children and of women being less developed and generally less active than in adult men, creatine rises to a higher level in blood and is excreted. Hyde [122] examined a group of men and women receiving 1 g of creatine daily. She observed that two young males accustomed to vigorous exercise retained more exogenous creatine than any of the other subjects. One of the women who was accustomed to vigorous exer-

References p. 302

cise each summer exceeded the other women and two young men of average musculature in creatine tolerance. The group included a young man with muscular dystrophy and a 59-year-old woman: they showed the lowest tolerances.

Creatinuria, more or less pronounced, is observed in starvation, in carbohydrate deprivation, in acidosis, in muscular dystrophies and a number of other pathological states (see Hunter [115], ch. VIII). The internal organs represent reservoirs for exogenous creatine. There are reservoirs of limited capacity such as the liver and the kidney, and more expansive reservoirs such as muscles. In the absence of creatine ingestion the deposition of creatine continues from filled reservoirs to unfilled ones [123]. On the other hand the synthesis of endogenous creatine is retarded in the presence of exogenous creatine. The repression of the arginine-glycine transamidinase which, as we shall see in a subsequent chapter, catalyses the formation, from glycine, of guanidinoacetic acid which becomes creatine through methylation, takes place through the influence of the creatine or the guanidinoacetic acid of the diet [124].

The presence of creatinine in urine was recognized by Liebig in 1847. In 1910, Shaffer and Reinoso [125] obtained, in protein-free filtrates of blood, a faintly positive Jaffé reaction (production of a red colour in an alkaline solution to which some aqueous picric acid had been added), as with dilute creatinine solutions.

Behre and Benedict [126], in 1922, criticized the specificity of the Jaffé reaction. Using rather unspecific techniques such as the comparison of the "chromogenic" material in blood and creatinine with respect to the action of hot alkalis (which destroy creatinine), to the adsorption by kaolin, etc., they concluded that "creatinine does not exist in blood in detectable quantities". After experiments of a similar nature, the same conclusion was reached by Gaebler [127] and by Bohn and Hahn [127].

The conclusion was not confirmed by other authors including Zacherl and Lieb, Ferro-Luzzi, Hayman, Johnston and Bender, and Danielson. (Literature in Miller and Dubos [129].)

Miller and Dubos [129], at the Rockefeller Institute for Medical Research, New York, approached the problem by using two bacterial enzymes, each having a high degree of specificity for creatinine. They concluded that true creatinine exists in human blood and constitutes almost all of the chromogen of the plasma. Since its discovery in urine in 1847, creatinine had naturally been considered as a waste product of creatine. (Literature in Hunter [115], p. 128.)

As stated in Chapter 10, in 1905 Folin [130] underlined the constancy of the daily excretion of creatinine in man. He considered it, not as a derivative of creatine, but "an index of a certain kind of protein metabolism occurring daily in any given individual" and conceived as "an essential part of the activity which distinguishes living cells from dead ones." According to the theory elaborated in 1914 by Folin and Denis [97], creatinine takes its origin "directly from protoplasm". In the context of the present discussion, the notion of the lack of convertibility of creatine into creatinine in animal organisms has been, since its formulation by Folin, contradicted repeatedly. (Literature in Hunter [115], p. 142.)

As stated above, the existence of creatinine in blood plasma was demonstrated by Miller and Dubos only in 1937. Since then, it has become clear that blood creatinine and urine creatinine are the result of a continuous production from muscle creatine phosphate by spontaneous non-enzymatic dephosphorylation (Borsook and Dubnoff [131]).

(e) Epistemological obstacles in early studies on creatine biosynthesis

As we shall see in Part V of the present *History*, recourse to isotopic methods led to the unravelling of the pathway of creatine biosynthesis which results from a process involving the interaction of three amino acids: glycine (the creatine backbone), arginine (the source of the transamidination forming guanidinoacetic acid, also called glycocyamine) and methionine (the source of the methyl) (Bloch and Schoenheimer [132], du Vigneaud [133]).

At the end of the period here considered — the 1930's — common belief reigned that creatine was a degradation product of arginine. Arginine had been identified as a product of protein hydrolysis near the turn of the centruy, when the constitution of creatine had already been identified, and it was logical to take

References p. 302

notice that both compounds contained a guanidine group: —C(=NH)—NH_2.

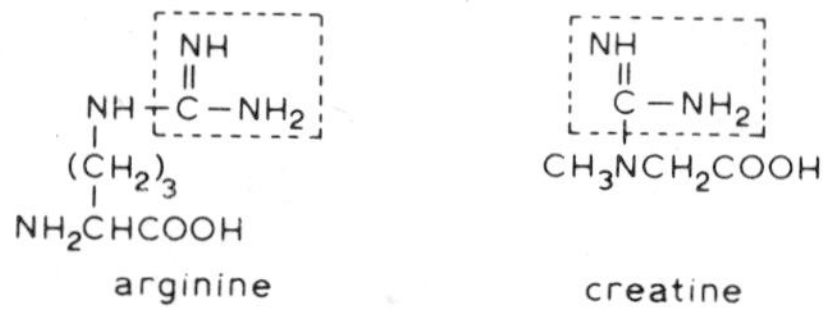

As we shall see, in 1927 Bergmann and Servas had already astutely recognized creatine as a derivative of glycine, and understood the participation of arginine as donor of an amidine group, but the pertinence of their views was to be recognized only in 1941, and confirmed by Bloch and Schoenheimer [132]. As stated above, a correct interpretation of the animal chemistry of the metabolic interrelations, among creatine phosphate, creatine and creatinine is a recent development, later than the end of the period considered. During this period, the subject matter considered allows for the identification of a number of epistemological and analytical obstacles, owing to which extensive researches, though correctly based chemically, remained fruitless. The experiments were accomplished with analytical procedures lacking specificity as well as sensitivity. The experimenters attempted to detect changes in the urinary excretion of creatine or creatinine, or changes in the creatine content of tissues. But the fluctuations in tissue composition and in excretion are very large. It was a common belief that creatine biosynthesis took place in muscle tissue, a belief that led to many irrelevant experiments. On the other hand, two epistemological obstacles exerted their effects until the end of the period considered. One was the notion of a lack of creatinine in blood plasma, which, as stated above, lasted from 1922 to 1937.

The other, even more adverse, notion was the concept of creatinine considered as a derivative, not of creatine, but of proteins. Introduced in 1905 by Folin, this concept lasted until approximately the same period of the late 1930's, when creatinine was finally recognized as a derivative of creatine, and as present in blood plasma. Until approximately the same data, a stumbling block remained against the evidence in favour of considering glycocyamine as an intermediate in creatine biosynthesis: its alleged absence from organisms. Glycocyamine was finally found in tissues and in urine by Weber [134,135].

It can be seen that the end of the period considered in the present volume not only coincides with the introduction of isotopic methods but also with the removal of a number of stumbling blocks as a consequence of progress in analytical methods.

(f) Creatine considered as a detoxicator of guanidine

This theory was proposed by Kutscher and Otori [136] in 1904. These authors, having observed that guanidine was liberated in the course of the autolysis of the pancreas, concluded that guanidine was a product of normal catabolism. In 1906, Achelis [137], obtained methylguanidine from normal urine, and he considered it almost certainly as a precursor of creatine. As early as 1905, Kutscher [138] had obtained methylguanidine from Liebig's meat extracts. Later, it was found in dried cod (Yoshimura and Kanai [139]) and in the fresh muscles of a number of animal species. (References in Hunter [115].)

W.F. Koch [140], in 1912, observed a high output of methylguanidine in the urine of parathyroidectomized dogs. Burns and Sharpe [141], in 1916, confirmed this. They observed not only that the guanidines of urine and blood were increased after parathyroidectomy, but that such an increase accompanied the idiopathic tetany of children. According to Paton and Findlay [142], both guanidine and methylguanidine produce symptoms similar to those observed in parathyroidectomized animals. During the same year, Paton and Findlay concluded that the symptoms of tetany are due to guanidine and methylguanidine, whose metabolism, according to their view, is controlled by the parathyroid. They found [143], in 1920, large amounts of methylguanidine in the urine of an adult case of tetany. In 1925, Paton [144] assigned to the parathyroid the function of regulating the change of the guanidines into creatine. A number of authors (literature listed in Hunter [115]) have tried in vain to demonstrate such a change.

(g) Influence of amino acids on creatine biosynthesis

It was on the basis of the nature of creatine, or methylguanidinoacetic acid, that the origin of its biosynthesis was looked for in the

References p. 302

only amino acid known to have a guanidine radical, namely arginine. In 1905, Czernecki [145], inspired by the known pathways of amino acid and fatty acid catabolism (see Part III), proposed a theory of creatine biosynthesis by the following reactions.

$$\underset{\text{arginine}}{HN{=}C(NH_2){-}NH{-}(CH_2)_3{-}CH(NH_2){-}COOH} \longrightarrow \underset{\gamma\text{-guanidine-butyric acid}}{HN{=}C(NH_2){-}NH{-}(CH_2)_3{-}COOH} \longrightarrow \underset{\text{guanidino-acetic acid (glycocyamine)}}{HN{=}C(NH_2){-}NH{-}CH_2{-}COOH} \longrightarrow \underset{\text{methylguanidino-acetic acid (creatine)}}{HN{=}C(NH_2){-}N(CH_3){-}CH_2{-}COOH}$$

Jaffé [146], and Lieben and Laszlo [147] observed no effect of arginine on the creatinine output in urine, creatinine being recognized at the time as a metabolite of creatine. No increase of creatine was observed after incubation with hashed muscle (Baumann and Marker [148]) or after perfusion through muscle (Baumann and Marker [148]).

Injecting arginine intravenously into rabbits, W.H. Thompson [149] observed a rise of muscle creatine. Positive effects of administering arginine on urine creatine were also observed in a number of animal species, mammals and birds (Thompson [149]; Gross and Steenbock [150]; Crowdle and Sherwin [151]). In dogs, Thompson [152] observed that the output of creatine is further augmented by the addition to the arginine (if hypodermically administered) of methyl citrate.

Concluding from the observations referred to above, Hunter [115] writes:

> "The experiments of Thompson were numerous and very carefully performed; and, if one accepts without question the evidence of the Jaffé reaction, leave no doubt that arginine administration is frequently, though not always, followed by an increase in creatine excretion. Whether this means a direct derivation of one from the other is not so certain" (p. 225).

In the 1920's it was generally considered that only inconclusive results were at hand about a direct conversion of arginine into creatine. The conversion of arginine into creatine involved, as the theory proposed, a passage through guanidinoacetic acid (glycocyamine). Glycocyamine was soon recognized as able to be transformed into creatine in vivo, but, as stated above, it was not considered, until 1935, as existing in organisms. In spite of a number of negative results (reported in Hunter [115]) the belief in the occurrence of this pathway became progressively generalized. That glycocyamine fed to rabbits led to increased excretion of creatinine and creatine was reported by Czernecki in the paper mentioned above. It was confirmed by Jaffé [146] at the laboratory of Medical Chemistry of Königsberg and by his assistant Dorner [153]. Experimenting on dogs and on ducks, W.H. Thompson [152] obtained the same result.

Gibson and Martin [154] experimented on a case of progressive pseudohypertrophic muscular dystrophy. This patient, who was completely intolerant to creatine, excreted in that form at least 36 per cent of 0.5 g of ingested glycocyamine.

Jaffé [146] Thompson [152] and Palladin and Wallenburger [155] reported that, in rabbits, the administration of glycocyamine produced an increase of creatine in muscles. These authors considered creatine as the result of the methylation of glycocyamine.

But, experimenting with growing chicks, Mellanby [156] was unable to demonstrate such an effect, neither did Baumann and Hines [157] with dogs. In a following paper, Baumann and Hines [158] described an increase of creatine after injecting glycocyamine, though they did not observe this effect in vitro when they added glycocyamine to hashed muscle, still considered at the time as the seat of creatine synthesis.

Bodansky [159,160] demonstrated an increase in creatine excretion and creatine concentration in rat muscle after administration of glycocyamine.

A very specific method, based on the use of a soil bacterium specifically destroying creatine and creatinine, was devised by Dubos and Miller [161]. Using this method, Borsook and Dubnoff [162], experimenting with liver slices of the cat, rabbit and rat, showed that glycocyamine is converted to creatine. To quote the authors (p. 560)

> "The difficulties and uncertainties which exist in conclusions resting on statistical analysis of small differences have therefore been overcome".

Borsook and Dubnoff found that, when methionine is present in the Ringer solution the creatine production is increased. The conversion of glycocyamine into creatine, an aspect of methylation,

Plate 171. William M. Bayliss.

exerted an influence in this field of metabolic studies. At the time, during the 1920's, methylation was attracting much attention from plant chemists. They believed it to be effected by the union of amino groups with formaldehyde, and reduction of the methylene-amino compounds produced. This view was still accepted in 1917 by W.H. Thompson [149], who suggested formaldehyde as the precursor of the methyl groups in creatine formation.

Creatine, a methylated amino acid derivative, was nevertheless suggested by W. Koch [163] to be metabolically related to choline, a methylated amino-alcohol. This view was developed by Riesser [164], who suggested that choline, or its oxidation product, betaine, could condense with urea, forming a guanidine group, in a synthesis similar to Volhard's synthesis of creatine from sarcosine and urea, Abderhalden and Buadze [165] reported in 1927 that the addition to hashed muscle or brain tissue of arginine and choline, together with arginase-containing liver "brei", resulted in an increase of creatine and creatinine. They suggest that this result is favourable to Riesser's theory of a formation of creatine with a participation of choline. In a second paper, Abderhalden and Möller [166] confirmed their observation and, avoiding conclusions about the sources of creatine, pointed to the possible participation of choline in its biosynthesis. Later, as we shall see, the utilization of methyl groups of choline for the synthesis of creatine was demonstrated by du Vigneaud and his colleagues (1941).

Cystine. That creatine could be derived from cystine "through the intermediate stages of taurine and amino-ethyl alcohol, followed by methylation, combination with urea and oxidation" was suggested by Harding and Young [167] on the basis of experiments (which have remained unpublished) in which they observed an increase of creatine excretion after feeding cystine to growing dogs. Gross and Steenbock, using pigs, also observed such an effect, but they considered it as due to an indirect effect of acidosis, resulting from the oxidation of the sulphur.

"Cystine feeding causes creatinuria only when the sulfuric acid formed by the oxidation of its sulfur is left unneutralized; when neutralized the creatinuria promptly disappears. Neutralization of acidity does not prevent the creatinuria called forth by casein or arginine feeding." (p. 35).

References p. 302

In the course of their observations, mentioned above, on patients suffering of pseudohypertrophic muscular dystrophy, Gibson and Martin found cystine without action on creatine elimination.

Histidine. As was stated in section 24 of Chapter 49, Ackroydt and Hopkins considered histidine as often interchangeable with arginine in metabolism. A number of authors (literature cited in Hunter [115]) have therefore tested, in vain, histidine with respect to a possible influence on creatine excretion.

Glycine and arginine. As stated above, arginine was long considered as providing the backbone of the molecule of creatine. Nevertheless, as early as 1927, Bergmann and Zervas [169] synthesized glycocyamine and creatine from arginine derivatives and glycine. Max Bergmann worked at that time at the Kaiser Wilhelm-Institut for leather research in Dresden. With Zervas, he accomplished the transfer of the amidine group of arginine to glycine, providing a model for the glycocyamine synthesis from arginine and glycine which was later to be identified by Bloch and Schoenheimer.

Brand, Harris, Sandberg and Ringer [170], in 1930, observed, that in patients suffering from muscular dystrophies, glycine caused an increase of creatinine elimination. Beard and Barnes [171] observed the same effect after feeding glycine to rats. Bodansky [59] found that, of the many alleged precursors of creatine besides guanidoacetic acid, glycine was the only one to increase the urinary creatine output "with any degree of regularity". But he considered the effect of glycine to be indirect — either stimulation of metabolism, increased renal function or washing out of creatine from tissues, etc. Brand and Harris [172] stated in 1933:

> "Our experiments indicate that the guanidine group of creatine is synthetic in origin and that glycine is involved in creatine formation".

This conclusion was based on increased creatine excretion after the feeding of glycine to patients suffering from progressive pseudohypertrophic muscular dystrophy (Brand, Harris, Sandberg and Ringer [170]) but also partly on the observation that administration of benzoic acid produced a drop in creatine excretion, presumably by removing glycine. But Bodansky [160] observed that, in the rat, the administration of benzoic acid did not diminish the output of creatine. Therefore, he did not consider glycine as involved in creatine formation.

At the end of the period considered, it was commonly thought that creatine was a product of arginine degradation.

8. Protein biosynthesis

The occurrence, in animals, of a biosynthesis of proteins from free amino acids was first considered by Cary [173] and by Blackwood [174]. The latter author observed a higher concentration of free amino acids in the afferent bloodstream to, than in the efferent one from, a site of protein synthesis. Block [175] proposed the existence of an "Anlage" of basic amino acids common to all proteins and acting as a kind of seed for the biosynthesis.

According to the theory of Alcock [176], based on the formaldehyde theory of photosynthesis, formaldehyde in plants is condensed with ammonia or nitrous acid to form a compound which by internal differentiation would form amino acids. According to this author the different proteins result from the association or elimination of amino acids, to or from a basic *Urprotein*. But the view that became current was that the biosynthesis of peptides and proteins was a result of a reversal of the reactions of protein hydrolysis, that is enzymatic synthesis of peptides and proteins. This theory, an aspect of the reverse enzymatic synthesis concept introduced by Van't Hoff in 1898, persisted for half a century.

References p. 302

REFERENCES

1 N.G. Coley, From Animal Chemistry to Biochemistry, Amersham, Bucks., 1973.
2 A.W. Hofman, in Faraday Lectures, 1869—1928, London, 1928.
3 A. Ure, London Med. Gazette (1841) cited after Williams [9], p. 12.
4 W. Keller, Ann. Chem., 43 (1842) 108.
5 F. Lieben, Geschichte der physiologischen Chemie, Leipzig and Vienna, 1935.
6 E. Baumann, Arch. Ges. Physiol., 12 (1876), 63, 69; 13 (1876) 285.
7 A. Kossel, Z. Physiol. Chem., 23 (1897) 1.
8 E. Baumann, Ber. Dtsch. Chem. Ges., 9 (1876) 54.
9 R.T. Williams, Detoxication Mechanisms. The Metabolism of Drugs and Allied Compounds, New York, 1947.
10 W. Henry, The Elements of Experimental Chemistry, 11th Ed., 2 vol., London, 1829.
11 J. von Mering and O. Musculus, Ber. Dtsch. Chem. Ges., 8 (1875) 662.
12 M. Jaffé, Z. Physiol. Chem., 2 (1878) 47.
13 O. Schmiedeberg and H. Meyer, Z. Physiol. Chem., 3 (1879) 422.
14 E. Sundwick, Jahresber. Fortschr. Tierch., 16 (1886) 76.
15 E. Fischer and O. Piloty, Ber. Dtsch. Chem. Ges., 24 (1891) 521.
16 E. Baumann and C. Preuse, Ber. Dtsch. Chem. Ges., 12 (1879) 806.
17 M. Jaffé, Ber. Dtsch. Chem. Ges., 12 (1879) 1092.
18 R. Neumeister, Lehrbuch der physiologischen Chemie, 2 vol., Jena, 1895.
19 J. Fruton, Molecules and Life. Historical Essays on the Interplay of Chemistry and Biology, New York, 1972.
20 A. Baeyer, Ber. Dtsch. Chem. Ges., 3 (1870) 63.
21 E. Baumann, Ueber die synthetischen Processe im Thierkörper, Berlin, 1878.
22 J. Fruton, Proceedings of the Conference on the Historical Development of Bioenergetics, Boston, Mass., 1975, p. 17.
23 M. Spaude, Eugen Albert Baumann (1846—1896). Zürich, 1973.
24 E. Pflüger, Arch. Ges. Physiol., 16 (1903) 1.
25 E. Pflüger, Arch. Ges. Physiol., 131 (1910) 302.
26 P.G. Stiles and G. Lusk, Am. J. Physiol., 9 (1903) 380.
27 G. Embden and H. Salomon, Beitr. Chem. Phys. Pathol., 5 (1904) 507.
28 Knopf, Arch. Exp. Pathol. Pharmacol., 49 (1903) 123.
29 A.I. Ringer and G. Lusk, Z. Physiol. Chem., 66 (1910) 106.
30 A.R. Mandel and G. Lusk, Am. J. Physiol., 16 (1906) 129.
31 A.I. Ringer, J. Biol. Chem., 15 (1913) 145.
32 H.D. Dakin and N.W. Janney, J. Biol. Chem., 15 (1913) 177.
33 M.P. Benoy and K.A. Elliott, Biochem. J., 31 (1937) 1268.
34 H. Weil-Malherbe, Biochem. J., 32 (1938) 2276.
35 C.F. Cori and G.T. Cori, J. Biol. Chem., 81 (1929) 389.
36 H.M. Edwards, Leçons sur la Physiologie et l'Anatomie Comparée de l'Homme et des Animaux, vol. 7, Paris, 1862.

37 J.B. Lawes and J.H. Gilbert, Phil. Mag., (4) 32 (1866) 439.
38 E. Meissl and F. Strohmer, Sitzber. Akad. Wiss. Wien, Math. Naturw. Klasse, 88 (1883) 205.
39 K.B. Lehmann and E. Voit, Z. Biol., 42 (1901) 619.
40 M. Rubner, cited by Lusk [41].
41 G. Lusk, The Elements of the Science of Nutrition, 4th Ed., Philadelphia, London, 1928.
42 H.J. Deuel, The Lipids. Their Chemistry and Biochemistry, vol. 2, New York, 1955.
43 A. Magnus-Levy, cited by Lusk [41].
44 R. Feulgen, K. Imhäuser and M. Behrens, Z. Physiol. Chem., 180 (1920) 161.
45 M. Pettenkofer and C. Voit, Z. Biol., 7 (1871) 433.
46 M. Rubner, Z. Biol., 21 (1885) 250.
47 E. Pflüger, Arch. Ges. Physiol., 52 (1892) 239.
48 M. Cremer, Münch. Med. Wochenschr., 44 (1897) 81.
49 A.F. Fourcroy, Ann. Chim. Phys., 3 (1789) 242.
50 M. Florkin, Un Prince, deux Préfets. Le Mouvement Scientifique et Médico-social au Pays de Liège sous le Règne du Despotisme Eclairé (1771—1830), Liège, 1957.
51 A.F. Fourcroy, Ann. Chim. Phys., 5 (1790) 154; 8 (1791) 17.
52 A.F. Fourcroy, Ann. Chim. Phys., 3 (1789) 120.
53 A.F. Fourcroy, Ann. Chim. Phys., 7 (1790) 146.
54 A.F. Fourcroy, Ann. Mus., 1 (1802) 93.
55 M.E. Chevreul, Ann. Chim. Phys., 95 (1815) 5.
56 A.B. Costa, Michel Eugène Chevreul, Pioneer of Organic Chemistry, Madison, 1962.
57 M.E. Chevreul, Ann. Chim. Phys., 2 (1816) 339.
58 M.E. Chevreul, Ann. Chim. Phys., 7 (1817) 155.
59 M.E. Chevreul, Recherches Chimiques sur les Corps Gras d'Origine Animale, Paris, 1823.
60 M.E. Chevreul, De la Méthode a Posteriori Expérimentale, Paris, 1870.
61 F. Reinitzer, Sitzber. Akad. Wiss. Wien, Math. Naturw. Klasse, Abt. I, 97 (1888) 167.
62 O. Diels and E. Abderhalden, Ber. Dtsch. Chem. Ges., 37 (1904) 3092.
63 J. Wislicenius and W. Moldenhauer, Ann. Chem., 146 (1868) 175.
64 A. Windaus, Z. Physiol. Chem., 213 (1932) 147.
65 O. Rosenheim and H. King, Nature, 130 (1932) 315.
66 J.D. Bernal, Chem. Ind., 51 (1932) 466.
67 O. Diels and W. Gädke, Ber. Dtsch. Chem. Ges., 60 (1927) 140.
68 H. Wieland and E. Dane, Z. Physiol., Chem., 210 (1932) 268.
69 S. Dezani, Arch. Farmacol. Sper., 16 (1913) 3.
70 S. Dezani and F. Cattoretti, Arch. Farmacol. Sper., 18 (1914) 3.
71 J.L. Gamble and K.D. Blackfan, J. Biol. Chem., 42 (1920) 401.
72 J.A. Gardner and F.W. Fox, Proc. Roy. Soc., B, 92 (1921) 358.
73 H. Dam, Biochem. Z., 215 (1929) 475.
74 R. Schoenheimer, Z. Physiol. Chem., 185 (1929) 119.

75 R.P. Hobson, Biochem. J., 29 (1935) 2023.
76 M. Tsujimoto, J. Soc. Chem. Ind., Japan, 9 (1906) 953.
77 I.M. Heilbron, E.D. Kamm and W.M. Owens, J. Chem. Soc., (1926) 1644.
78 H.J. Channon, Biochem. J., 20 (1926) 400.
79 P. Karrer and A. Helfenstein, Helv. Chim. Acta, 14 (1931) 78.
80 L. Ruzicka, Proc. Chem. Soc. (1959) 341.
81 R. Robinson, J. Soc. Chem. Ind., 53 (1934) 1062.
82 J.L.W. Thudichum, Report of the Medical Officer to the Privy Council for 1867, vol. X, Appendix 7, p. 152.
83 F. Hoppe-Seyler, Med.-Chem. Unters., 4 (1871) 531.
84 M. Nencki and J. Zaleski, Z.Physiol. Chem., 34 (1901) 997.
85 K. Dobriner and C.P. Rhoads, Physiol. Rev., 20 (1940) 416.
86 H. Fischer and H. Orth, Die Chemie des Pyrrols, 2 vol., Leipzig, 1934.
87 H. Fischer and K. Zeile, Ann. Chem., 468 (1929) 98.
88 C. Rimington, C.R. Lab. Carlsberg, 22 (1938) 454.
89 W.J. Turner, J. Lab. Clin. Med., 26 (1940) 323.
90 A.P. Briggs, J. Biol. Chem., 53 (1922) 13.
91 P. Eggleton and G.P. Eggleton, Biochem. J., 21 (1927) 190.
92 C. Fiske and Y. Subbarow, J. Biol. Chem., 66 (1925) 375.
93 C. Fiske and Y. Subbarow, Science, 65 (1927) 401.
94 C. Fiske and Y. Subbarow, Science, 67 (1928) 169.
95 C. Fiske and Y. Subbarow, J. Biol. Chem., 81 (1929) 629.
96 F. Urano, Beitr. Chem. Phys. Pathol., 9 (1906) 104.
97 O. Folin and W. Denis, J. Biol. Chem., 17 (1914) 493.
98 G. Embden and E. Adler, Z. Physiol. Chem., 118 (1922) 1.
99 O.W. Tiegs, Austral. J. Exp. Biol. Med., 2 (1924—1925) 1.
100 G. Stella, J. Physiol., 66 (1928) 19.
101 H. Lehmann, Biochem. Z., 281 (1935) 271.
102 I. Torres, Biochem. Z., 283 (1936) 128.
103 M.E. Chevreul, J. Pharm., 21 (1835) 231.
104 J. Liebig, Compt. Rend., 24 (1847) 69, 195.
105 J. Liebig, Ann. Chim. Phys., 62 (1847) 257.
106 D.S. Price, Quart. J. Chem. Soc., 3 (1851) 229.
107 W. Gregory, Ann. Chem., 64 (1848) 100.
108 Schlossberger, Ann. Chem., 66 (1848) 80.
109 F. Grohé, Ann. Chem., 85 (1853) 233.
110 Schlossberger, Ann. Chem., 49 (1844) 341.
111 V. Dessaignes, Compt. Rend. 38 (1854) 839.
112 A. Strecker, Lehrbuch der organischen Chemie, 5te Aufl., 1867, p. 586.
113 A. Strecker, Compt. Rend., 52 (1861) 1210.
114 J. Volhard, Sitzungsber. königl. Bayer. Akad. Wiss., 1868 (II) 472.
115 A. Hunter, Creatine and Creatinine, London, 1928.
116 E. Erlenmeyer, Ann. Chem., 146 (1868) 259.
117 J. Horbaczewski, Wiener Med. Jahrb., (1855) 459.
118 W. Paulmann, Arch. Pharm., 232 (1894) 601.
119 W.C. Rose, J. Biol. Chem., 10 (1911—1912) 265.

120 O. Folin and W. Denis, J. Biol. Chem., 11 (1912) 253.
121 R.A. Krause, Quart. J. Exp. Physiol., 4 (1911) 293.
122 E. Hyde, J. Biol. Chem., 142 (1942) 301.
123 H.D. Hoberman, E.A.H. Sims and J.H. Peters, J. Biol. Chem., 172 (1948) 45.
124 J.B. Walker, J. Biol. Chem., 236 (1961) 493.
125 P.A. Shaffer and E.A. Reinoso, J.Biol. Chem., 7 (1909—1910) xxx.
126 J.A. Behre and J.R. Benedict, J. Biol. Chem., 52 (122) 11.
127 O.H. Gaebler, J. Biol. Chem., 89 (1930) 454; 117 (1937) 397.
128 H. Bohn and F. Hahn, Z. Klin. Med., 125 (1933) 458.
129 B.F. Miller and R. Dubos, J. Biol. Chem., 121 (1937) 447.
130 O. Folin, Am. J. Physiol., 13 (1905) 117.
131 H. Borsook and J.W. Dubnoff, J. Biol. Chem., 168 (1947) 493.
132 K. Bloch and R. Schoenheimer, J. Biol. Chem., 138 (1941) 167.
133 V. du Vigneaud, A Trail of Research, Ithaca, 1952.
134 C.J. Weber, J. Biol. Chem., 109 (1935) XCVI.
135 C.J. Weber, J. Biol. Chem., 114 (1936) CVII.
136 F. Kutscher and J. Otori, Z. Physiol. Chem., 43 (1904) 93.
137 W. Achelis, Z. Physiol. Chem., 50 (1906) 10.
138 F. Kustscher, Unters. Nahr. Genussm., 10 (1905) 528.
139 K. Yoshimura and M. Kanai, Z. Physiol. Chem., 88 (1913) 346.
140 W.F. Koch, J. Biol. Chem., 12 (1912) 313.
141 D. Burns and J.S. Sharpe, Quart. J. Exp. Physiol., 10 (1916) 345.
142 D.N. Paton and L. Findlay, Quart. J. Exp. Physiol., 10 (1916) 315.
143 D.N. Paton and L. Findlay, Quart. J. Exp. Physiol., 13 (1920) 433.
144 D.N. Paton, Glasgow Med. J., Dec. 1925.
145 W. Czernecki, Z. Physiol. Chem., 44 (1905) 294.
146 M. Jaffé, Z. Physiol. Chem., 48 (1906) 430.
147 F. Lieben and D. Laszlo, Biochem. Z., 176 (1926) 403.
148 L. Baumann and J. Marker, J. Biol. Chem., 22 (1915) 49.
149 W.H. Thompson, J. Physiol., 51 (1917) 111.
150 E.G. Gross and H. Steenbock, J. Biol. Chem., 47 (1921) 33.
151 H.J. Crowdle and C.P. Sherwin, J. Biol. Chem., 55 (1923) 363.
152 W.H. Thompson, J. Physiol., 51 (1917) 347.
153 G. Dorner, Z. Physiol. Chem., 52 (1907) 225.
154 R.B. Gibson and F.T. Martin, J. Biol. Chem., 49 (1921) 319.
155 A. Palladin and L. Wallenburger, C.R. Soc. Biol., 78 (1915) 111.
156 E. Mellanby, J. Physiol., 36 (1908) 447.
157 L. Baumann and H.M. Hines, J. Biol. Chem., 24 (1916) 439.
158 L. Baumann and H.M. Hines, J. Biol. Chem., 31 (1917) 549.
159 M. Bodansky, J. Biol. Chem., 112 (1936) 615.
160 M. Bodansky, J. Biol. Chem., 115 (1936) 641.
161 R. Dubas and B.F. Miller, J. Biol. Chem., 121 (1937) 429.
162 H. Borsook and J.W. Dubnoff, J. Biol. Chem., 32 (1940) 559.
163 W. Koch, Am. J. Physiol., 15 (1905) 15.
164 O. Riesser, Z. Physiol. Chem., 86 (1913) 415; 90 (1914) 221.
165 E. Abderhalden and S. Buadze, Z. Physiol. Chem., 164 (1927) 280.

166 E. Abderhalden and P. Möller, Z. Physiol. Chem., 170 (1927) 212.
167 V.J. Harding and E.G. Young, J. Biol. Chem., 41 (1920) XXXVI.
168 E.G. Gross and H. Steenbock, J. Biol. Chem., 47 (1921) 33.
169 M.Bergmann and L. Zervas, Z. Physiol. Chem., 172 (1927) 277.
170 E. Brand, M.M. Harris, M. Sandberg and A.I. Ringer, Am. J. Physiol., 90 (1930) 296.
171 H.H. Beard and B.O. Barnes, J. Biol. Chem., 94 (1931—1932) 49.
172 E. Brand and M.M Harris, Science, 77 (1933) 589.
173 C.A. Cary, J. Biol. Chem., 67 (1926) XXXIX.
174 J.H. Blackwood, Biochem. J., 26 (1932) 772.
175 R.J. Block, J. Biol. Chem., 104 (1934) 347; 105 (1934) 45.
176 R.S. Alcock, Physiol. Rev., 16 (1936) 1.

Chapter 51

Reversible Zymo-hydrolysis

1. Reversible zymo-hydrolysis of carbohydrates and glucosides

The expression "reversible zymo-hydrolysis" was introduced by A.C. (Croft) Hill [1] in 1898 to designate the alleged promotion of a synthesis of maltose from glucose in the presence of maltase. Hill had been inspired by a paper of Van't Hoff [2] who had considered the possibility of mass action reversal in catalysed reactions. The conclusion of Hill was not accepted by all. Oppenheimer [3], for instance, did not consider the evidence as convincing. To account for the small amounts of maltase obtained, Hill suggested that the diversion of starch in insoluble form in plants, or of glycogen in animals, permitted the synthesis to proceed.

In a paper of 1905, Visser [4] compared the results of the hydrolysis of saccharose by invertase and by acid. He pointed out that the hydrolysis in the presence of invertase appeared to be complete. When acted upon by invertase until no further change took place, a 0.25 N solution of saccharose gave a rotation of —3.26°. But when saccharose was acted upon by acid, the final rotation was —3.42°, which is what the reading should be if the solution contained only glucose and fructose. Visser observed that a solution containing equal amounts of glucose and of fructose had an initial rotation of —12.46°. When acted upon *for two months* by invertase the rotation had fallen to —12.29°. A change of such degree means that the equilibrium position is reached when about 99 per cent of the saccharose is hydrolysed. It also means that if a solution of glucose and fructose is acted upon by invertase, about 1 per cent should be converted into saccharose. The equilibrium position is given by the ratio of the velocity of

Plate 172. Elie Emile Bourquelot.

the hydrolysis to that of the synthetic process. Visser found that for 0.5 N scccharose, the equilibrium constant (i.e. the ratio of the two velocity constants) was very nearly 50. Visser [4] (p. 301) concluded that, since six days were required to attain equilibrium when saccharose of this concentration was acted upon by invertase, *about ten months* (i.e. fifty times six days) would be needed for the reverse reaction.

It may appear surprising that a synthesis of such small an extent, taking such long a time, could have been considered as biologically important.

One of the converts of "reversible zymo-hydrolysis", Bayliss [5], has explained that it "would be an error" not to take this kind of biosynthesis in consideration. Bayliss thought that in spite of their smallness, the amounts synthesized were of great importance, as not only could the product be deposited in an insoluble form but

"it may be removed from the reacting system by any other means, such as diffusion into blood-current or elsewhere, or taken up in some other independent reaction" (p. 54).

As stated by Wasteneys and Borsook [6].

"The essence of the theory is the applicability of the Law of Mass Action to enzyme reactions, and their reversibility consequent upon the assumed catalytic nature of the reaction".

From the theory it was assumed that a sufficient reduction of the concentration of water would result in the reaction's proceeding towards the direction of synthesis.

Bayliss, in his *Principles of General Physiology* [7], a widely read book, considered the case of an alcohol-water-acid-ester system and he calculated the equilibrium constant as follows:

$$\frac{(\text{alchol})(\text{acid})}{(\text{ester})(\text{water})} = \frac{K_1}{K_2} = K$$

To quote Bayliss [7] (p. 325)

"Put in this form, we see that if we increase one component, the result must be to decrease its fellow, since the value of the fraction must remain unaltered. Suppose we increase water, the value of this fraction can only be kept constant either by increasing (alcohol)(acid) or by decreasing (ester). In point of fact, of course, the two are identical, since one cannot take place without

Plate 173. Marc Bridel.

the other. The results of excess of water should be, therefore, to increase the hydrolytic reaction of the system, as found by experiment. The conclusion to be drawn from this fact is that, in order to obtain much indication of the synthetic aspect of the enzyme action, the concentration of water must be diminished as far as possible".

A vast literature has accumulated on enzymatic synthesis in the field of carbohydrates and glucosides. Following the idea of reducing the concentration of water in the reaction medium, Robertson, Irvine and Dobson [8] described a production of sucrose, to the extent of 6 per cent, when they allowed the "sucroclastic enzymes" of *Beta vulgaris* to act on a concentrated solution of invert sugar. But the same authors failed, during the period of sugar storage, to find these enzymes in the roots.

Bourquelot and Bridel [9], in Paris, with the aid of emulsin, made a number of the β-glucosides of various primary alcohols.

From these different contributions it was concluded that there was no cogent evidence that enzymes produce by synthesis any substances different from those which they hydrolyse (literature cited in Bayliss [5]).

2. Synthesis in the presence of lipase

This was first described by Kastle and Loevenhart [10]. Since the ester, ethyl butyrate, produced by the action of lipase on a mixture of ethyl alcohol and butyric acid, has a characteristic odour, it was easy to observe the production of ethyl butyrate as reported by the authors.

"When a fresh aqueous extract of pancreas is treated with a mixture of dilute butyric acid (0.1 to 0.05 N) and ethyl alcohol (sufficient to bring the whole to 1.5 per cent), the very characteristic odour of ethyl butyrate soon develops even at the ordinary temperature and in the presence of antiseptics whereas if the pancreatic extract is first boiled, the mixture never develops the odour of the ester". (Quoted after Bayliss [5], p. 55.)

It was also observed that when ethyl butyrate was hydrolysed in the presence of lipase, an equilibrium was attained and the reaction was never complete.

Because pancreatic lipase not only catalyses the hydrolysis of simple esters, but also of higher fats, the synthetic production of higher esters of glycerol was expected, and it was actually described

References p. 323

Plate 174. Maurice Hanriot.

by Hanriot [11] who, in the presence of lipase, obtained monobutyrin (butyric ester of glycerol), and by Pottevin [12], who obtained mono- and tri-olein.

As the existence of the enzymatic synthesis of glycerides was generally admitted, it gave rise to physiological interpretations. For instance, Loevenhart [13], in 1902, explained on such a basis the re-synthesis of fat globules in the intestinal mucosa from products of hydrolysis of glycerides in the intestinal lumen. This notion was considered as supported by the finding of lipase in the intestinal mucosa of the pig as well as in liver and other sites of lipogenesis.

How ingrained the theory of fat biosynthesis by reversal of the hydrolytic actions had become is shown by the general agreement to the notion that bile salts, known to accelerate hydrolysis, would be expected to accelerate the synthetic process, and the alleged demonstration by Hansik [14], of this effect. The agreement, however, was not universal. Determining the amount of fats contained in various tissues and the lipase content of those, Bradley [15] concluded that, as there is no parallelism between the two, the view that the fat is synthesized by the enzyme should be repudiated. But the supporters of synthesis by reversal of enzymatic action answered this criticism by claiming that "reversible zymo-hydrolysis" may be carried on in one place and the product stored or excreted in another place. As stated by Bayliss [5] (p. 16)

"In fact, it is quite conceivable that this may be the more effective method, owing to the removal of the products allowing continuous synthesis to proceed".

3. Plasteins

The concept of a formation of peptide bonds by "reversible zymohydrolysis" has long been considered as supported by the formation of so-called "plasteins".

(a) Definition

In a book published in 1886 (i.e. before the publication of A.C. Hill's paper), Danilevski [16] described the production of a pre-

References p. 323

Plate 175. Alexander Jakovlevich Danilevsky.

cipitate by adding rennet preparations to concentrated solutions of Witte's peptone. In 1901, Savjalov [17] repeated the experiment of Danilevski, and obtained a precipitate which he called *plastein*.

Lavrov and Salaskin [18] were not convinced that the plasteins were of protein nature and they considered them as mixtures of "proteoses". But Savjalov [19], in 1907, restated his position regarding the concept of plasteins as results of a true synthetic process, evidence of a reversible action of pepsin.

A large literature accumulated for half a century on the precipitates (plasteins) obtained as a consequence of the addition of a proteolytic enzyme to a concentrated solution of the product of hydrolysis by pepsin. The enzyme added varied in different experiments. Plastein formation after adding pepsin was described by Lavrov and Salaskin [18], Bayer [20], Robertson [21], Henriques and Gjaldbäk [22], Borsook and Wasteneys [23].

Not only was a formation of plastein observed by adding pepsin to concentrated pepsin digests (literature cited by Wasteneys and Borsook [6]) but also after the addition to such concentrated pepsin digests of other proteolytic enzyme preparations, for instance papayotin (Kurajeff [24]), activated pancreatic juice (Delezenne and Mouton [25]) or commercial trypsin (Henriques and Gjaldbäk [22]; Wasteneys and Borsook [26]). On the other hand, the claim of a plastein formation after the addition of trypsin to a concentrated trypsin digest, claimed by Marston [27], was not confirmed by Henriques and Gjaldbäk [28] or by Wasteneys and Borsook [29] who confirmed that plastein formation, possible by the action of proteolytic enzymes on mixtures of polypeptides, could not be observed by the same action on mixtures of amino acids.

The sophistication of experimentation on plastein formation increased when the enzymatic preparation added became the product of tissue autolysis (Nürnberg [30]; Grossmann [31]; Taylor [32,33]). Abderhalden [34] allowed a tissue to autolyse for three months. He then concentrated it and added a fresh extract of the same kind of tissue. The mixture was left for five months in conditions claimed to ensure sterility. Reversibility was described for liver, thyroid, lung and kidney but no reversal was described when an autolysate of yeast was used or when the fresh extract of tissue added was obtained from another tissue than that used to prepare the hydrolysate. This experiment seemed impres-

References p. 323

sive and in favour of protein formation. The author showed the appearance of heat-coagulable material, giving a positive biuret test. In the reacting mixture, as had already been observed by Henriques and Gjaldbäk [22], the concentration of free amino nitrogen was lowered after plastein formation, as well as the intensity of the colour test for free tryptophan. This intensity was restored after the hydrolysis of the synthesized plastein. On the other hand, as noted by Wasteneys and Borsook [6], the paucity of yield as well as the length of time of the experiment jeopardized the belief in its biological significance.

(b) Was the plastein precipitate of proteinic nature?

At the time of experimentation on plasteins, the nature of the products of the action of proteolytic enzymes on proteins was still under discussion. Mialhe [35] considered that the soluble products of the digestion of all proteins were identical, and he called the resulting substance "albuminose." Lehmann [36], in the second volume of his *Lehrbuch* (p. 317), stated that he had observed differences among the products of proteolytic enzymes on proteins and he renamed them "peptones." As we shall see, the nature of the products of the action of proteolytic enzymes remained obscure for a long time and the significance of such intermediates as "albumoses" (or "albuminoses", "peptones", "proteoses" etc.) for a long time remained undefined in terms of mixtures of polypeptides.

In most experiments on "plastein" formation recorded in the literature it was assumed that the plastein precipitate was of a more complex nature than the protein derivatives present in the hydrolysed mixture at the start. Several arguments supported this notion. The precipitate was only formed in concentrated solutions of the hydrolysate, and no precipitate was formed if the enzyme were not added. In his paper of 1907, Savjalov [19] had stated that plasteins suspended in acid solutions were digested by pepsin. This was confirmed by Wasteneys and Borsook [37] (p. 121) who found the rate of hydrolysis the same as with native proteins.

(c) Are plasteins real synthetic products?

One of the arguments given in support of this concept was the decrease in free amino nitrogen recorded by Henriques and Gjaldbäk [22]. This was contested by Oda [38] in 1926; but finally confirmed by Rona and Oelkers [39]. From fractional analysis of a digest in which different amounts of plasteins had been formed, Wasteneys and Borsook [40] concluded that all fractions designated by the terminology of the time (proteoses, peptones, subpeptones) were involved in the formation of plastein while no plastein resulted from the action of trypsin on tryptic digests. Lieben [41] considered Borsook's paper, together with that of Rona and Oelkers [39] as having established the existence of the "merkwürdig" plastein formation resulting from the addition of pepsin to pepsin digests.

(d) Properties of plasteins

Regardless of the nature of the proteins that were the sources of the digest which was concentrated at the start, the plasteins showed similar properties. The proteoses and peptones that became associated as plasteins were associated by chance meeting with the enzyme and did not have a constant composition.

Levene and Van Slyke [42,43] compared the amino acid compositions of fibrin and of the plastein obtained from a peptic digest of fibrin (Witte's peptone). The conclusion was that the plasteins, from which they isolated at least 13 amino acids, were nearer to polypeptides than to the proteins. More recently, Mensorow [44] has provided convincing data in favour of the polypeptidic character of plasteins.

(e) Assumed association with respiratory processes

The concept of plastein formation was tentatively associated with respiratory processes. Voegtlin and his collaborators, having hydrolysed fibrin, or other proteins, in the presence of papain, claimed to have reversed the hydrolysis by an intensive oxygenation. The interpretation was that "oxydated papain", which they

References p. 323

Plate 176. Valdemar Henriques.

considered to be formed, catalysed the reaction that was the inverse of hydrolysis [45]. But it is clear that no reaction can be reversed by changing the catalyst, unless the Mass Action conditions are altered.

Linderstrøm-Lang and his collaborators failed to confirm the observations of Voegtlin, and they pointed to methodological errors in his experiments. In the course of their observations, Strain and Linderstrøm-Lang [46] observed, when stirring hydrolysis products of wool (rich in cysteine) by papain, in oxygen in the presence of papain, the formation of substances precipitable by trichloroacetic acid, but they proved that no formation of peptide bonds accompanied such "syntheses". They showed that the number of SH groups was reduced, which they explained by the aggregation through such SH groups of polypeptides forming, in the presence of oxygen, S—S bridges with the production of larger and less soluble peptide aggregates.

4. Impossibility of formation of peptide bonds by reversal of Mass Action

The work of Borsook and Wasteneys on the synthesis of proteins was accomplished in Toronto. The idea that was the background of that work was the possibility of synthesis by Mass Action reversal, and it was considered as confirmed by the "plastein" formation resulting from the action of pepsin on concentrated *peptides* (not on amino acids) resulting from protein digestion (though, as stated above, a formation of S—S bridges had been suggested). Borsook moved to the California Institute of Technology, Pasadena, in 1929. While plastein formation arose from concentrated solutions of peptides, it had been impossible to obtain plasteins from tryptic digests. As noted above, this important conclusion had been fully documented by Wasteneys and Borsook [26].

As stated by Borsook [47], at about the time of his arrival at Cal Tech,

"To improve my knowledge of physical chemistry, I studied Noyes and Sherrill's text which the undergraduates were using. From the section on thermodynamics the question which arose in my mind was whether the failure to get protein synthesis from concentrated amino acid solutions was because the equilibrium of tryptic hydrolysis, for example, was away over at practically com-

Plate 177. Henry Borsook.

plete hydrolysis, i.e. the free energy change in hydrolysis was so large that one could not get synthesis by any feasible degree of concentration of amino acids. It follows that energy is needed to be put in the system, i.e. it had to be a coupled reaction." (p. 278).

Direct measurements of the free energy of formation of peptide bonds were impossible. Borsook then learned of the work of Parks and his pupil Huffman at Stanford. As Borsook states, Parks and Huffman

"extended Giauque's work on the determination of the entropy of compounds by integrating the specific heat from 0°C to any desired temperature. Parks and Huffman had a program of such measurements on organic compounds." (p. 278).

Borsook succeeded in bringing Huffman to Pasadena to do the work on biologically important compounds [48].

"To get the free energies, specific heats and heats of combustion were necessary. Then, with solubility and ionization data, the free energy in solution was calculated. In this way we learned that the free energy of formation of the peptide bond was 2500—3000 calories, that reversal by mass action was out of the question, synthesis had to be coupled with free energy donating reaction." (p. 278).

This disposed of the alleged peptide bond formation in plasteins, deduced from experiments accomplished with unpurified peptone preparations and very crude enzymatic extracts, and in which no proof had been provided of a formation of peptide bonds.

5. Physiologically reversed reactions of biosynthesis as distinguished from physicochemically reversible reactions

The first peptide bond synthesis shown to be coupled with a free-energy-donating reaction was the synthesis of hippuric acid from benzoic acid and glycine. Borsook and Dubnoff [49], having shown that it is attended by a gain in free energy, studied the reaction as taking place in liver slices and concluded that the synthesis was completely inhibited by 0.001 M KCN and by toluene, showing that cell respiration was essential for the synthesis. But the energetic aspect was not the only feature emphasized. The authors found that liver slices treated with cyanide did not hydrolyse

References p. 323

hippuric acid added to the Ringer solution. To quote Borsook and Dubnoff [49]:

"If the synthesis of hippuric acid were simply the reverse of hydrolysis brought about by a shift of the equilibrium through coupling with an energy-donating reaction, it may be expected that when this coupling is broken by a respiratory poison hydrolysis of any hippuric acid present would ensue. This would have been analogous to the hydrolysis of protein in autolysis. The finding that there is no hydrolysis in liver slices poisoned with cyanide indicated that the synthesis of hippuric acid under the conditions we have employed, and probably also under those in vivo, is not simply the reverse of hydrolysis" (p. 321).

"The hydrolysis and synthesis of hippuric acid in the dog is an example of a reaction which is physiologically but not physicochemically "reversible", as reversibility is ordinarily understood." (p. 322).

We see here the expression of the concept that however energetically satisfied the reverse of hydrolysis could be considered, the biosynthetic pathway is not necessarily what the theory of "reversible zymo-hydrolysis" suggested. That the chemical pathway of the biosynthesis is not indentical with the pathway taken in the degradation had already been illustrated by the degradation of urea into glyoxylic acid and ammonia in the presence of urease and by the synthesis of urea through the ornithine cycle, unravelled by Krebs and Henseleit since 1932.

But a decisive argument came in the form of the description by Cori, Cori and Schmidt, in 1939 of the biosynthesis of polysaccharides by the reversible reaction:

$$Gl_1Gl_2Gl_3 \ldots Gl_n + \text{glucose-1-P} \rightleftharpoons Gl_1Gl_2Gl_3 \ldots Gl_{n+1} + \text{phosphate}\ .$$

in the presence of a specific phorphorylase *.

Whereas before 1940, the generality of biosynthesis by reversible zymo-hydrolysis formulated in 1898 was accepted, many more examples besides those enumerated above reinforced, after 1940, the principle that the pathway of biosynthesis was not usually identical with the degradation pathway.

---

* See Notes added in proof, p. 362.

REFERENCES

1 A.C. Hill, J. Chem. Soc., 73 (1898) 634.
2 J.H. Van't Hoff, Z. Anorg. Chem., 18 (1898) 1.
3 C. Oppenheimer, Die Fermente und ihre Wirkung, Leipzig, 1900.
4 A.W. Visser, Z. Phys. Chem., 52 (1905) 257.
5 W.M. Bayliss, The Nature of Enzyme Action, London, 1919.
6 H. Wasteneys and H. Borsook, Physiol. Rev., 10 (1930) 110.
7 W.M. Bayliss, Principles of General Physiology, 4th ed., London, 1924.
8 R.A. Robertson, J.C. Irvine and M.E. Dobson, Biochem. J., 4 (1909) 258.
9 E. Bourquelot and M. Bridel, Ann. Chim. Phys., 28 (1913) 145.
10 J.H. Kastle and A.S. Loevenhart, Am. Chem. J., 24 (1900) 491.
11 M. Hanriot, C.R. Soc. Biol., 53 (1901) 70.
12 H. Pottevin, Bull. Soc. Chim., 35 (1906) 693.
13 A.S. Loevenhart, Am. J. Physiol., 6 (1902) 331.
14 A. Hamsik, Z. Physiol. Chem., 65 (1910) 232.
15 H.C. Bradley, J. Biol. Chem., 13 (1913) 407.
16 B. Danilevski, The Organoplastic Forces of the Organism (in Russian), Charkoff, 1886 (cited after Bayliss [5], p. 173).
17 W.W. Savjalov, Arch. Ges. Physiol., 75 (1901) 171.
18 M. Lavrov and S. Salaskin, Z. Physiol. Chem., 36 (1902) 277.
19 W.W. Savjalov, Z. Physiol Chem., 54 (1907) 119.
20 H. Bayer, Beitr. Chem. Physiol. Pathol., 4 (1903) 554.
21 T.B. Robertson, J. Biol. Chem., 3 (1907) 95.
22 V. Henriques and I.K. Gjaldbäk, Z. Physiol. Chem. 71 (1911) 485.
23 H. Borsook and H. Wasteneys, J. Biol. Chem., 62 (1924—1925) 633.
24 D. Kurajeff, Beitr. Chem. Physiol. Pathol., 1 (1901—1902) 12.
25 C. Delezenne and H. Mouton, C.R. Soc. Biol., 63 (1907) 277.
26 H. Wasteneys and H. Borsook, J. Biol. Chem., 63 (1925) 575.
27 H.R. Marston, Austr. J. Exp. Biol. Med., 3 (1926) 233.
28 V. Henriques and I.K. Gjaldbäk, Z. Physiol. Chem., 81 (1912) 439.
29 H. Wasteneys and H. Borsook, Z. Physiol. Chem., 62 (1924—1925) 15, 675.
30 A. Nürnberg, Beitr. Chem. Phys. Pathol., 4 (1903) 543.
31 J. Grossmann, Beitr. Chem. Phys. Pathol. 4 (1905) 194.
32 A.E. Taylor, J. Biol. Chem., 3 (1907) 87.
33 A.E. Taylor, J. Biol. Chem., 5 (1909) 381.
34 E. Abderhalden, Fermentforschung, 1 (1916) 47.
35 L. Mialhe, Mémoire sur la Digestion et l'Assimilation des Matières Albuminoïdes, Paris, 1847.
36 C.G. Lehman, Lehrbuch der physiologischen Chemie, 2nd Ed., Leipzig, 1853.
37 H. Borsook, Physiol. Rev., 30 (1950) 206.
38 T. Oda, J. Biochem. (Jap.), 6 (1926) 77.
39 P. Rona and H.A. Oelkers, Biochem. J., 203 (1928) 298.
40 H. Wasteneys and H. Borsook, J. Biol. Chem., 62 (1924—1925) 5.

41 F. Lieben, Geschichte der physiologischen Chemie, Leipzig and Vienna, 1935, p. 390.
42 P.A. Levene and D.D. Van Slyke, Biochem. Z., 13 (1908) 458.
43 P.A. Levene and D.D. Van Slyke, Biochem. Z., 16 (1909) 203.
44 I.G. Mensorow, Enzymologia, 10 (1941) 127.
45 M.E. Maver and C. Voegtlin, Enzymolgia, 6 (1939) 219.
46 H.H. Strain and K. Linderstrøm-Lang, C.R. Lab. Carlsberg, 23 (1940) 167.
47 H. Borsook, Proc. Conference on the Historical Development of Bioenergetics, Boston, Mass. 1975, p. 278.
48 H. Borsook and H.M. Huffman, in C.L.A. Schmidt (Ed.), Chemistry of the Amino Acids and Proteins, Springfield, Ill., 1938.
49 H. Borsook and W. Dubnoff, J. Biol. Chem., 132 (1940) 307.
50 G.T. Cori, C.F. Cori and G. Schmidt, J. Biol. Chem., 131 (1939) 397.
51 G.T. Cori, C.F. Cori and G. Schmidt, J. Biol. Chem., 135 (1940) 733.
52 D. Nachmansohn, R.T. Cox, C.W. Coates and A.L. Machado, J. Neurophysiol., 6 (1943) 383.
53 H.A. Krebs, Enzymology, 12 (1946—1948) 88.

Chapter 52

First Approaches to the Biosynthesis of Amino Acids

1. Reducing amination of keto acids

It was generally considered in the 1920's that the non-essential amino acids were synthesized from α-keto acids by amination. This view originated from experiments of Knoop [1,2] who fed keto acids to animals and obtained an excretion of amino acids. Embden and Schmitz [3], as well as Dakin and Dudley [4], reached the same conclusion from experiments on perfused isolated liver. They found that alanine, phenylalanine and tyrosine might be synthesized in the liver from the ammonium salts of the corresponding ketonic acid. Embden and Schmitz [3] observed that ammonium lactate might yield alanine. Though at the time pyruvate had not yet been isolated it was considered that the catabolism of alanine or the pathway of lactic acid to alanine took place with the intermediate formation of pyruvic acid. Later, Neber [5] observed a formation of amino acids after adding pyruvate and ammonia to liver slices.

The theories proposed to account for amination of α-keto acids have, in several of their aspects, been influenced by the theory of a reversal by Mass Action, of reactions catalysed by enzymes, as stated in Chapter 51.

Discussing the mechanism of the formation of amino acids by the amination of keto acids, Knoop considered two hypotheses. In the first, he considered the biosynthesis as a reversal of the process of oxidative deamination. As a result of the contributions of Neubauer [6,7], or Knoop [8] and of Krebs [9,10], the scheme currently accepted with respect to oxidative deamination read as follows

$$\mathrm{H-\overset{\displaystyle R}{\underset{\displaystyle COOH}{C}}-NH_2} \xrightarrow{+\,O_2} \mathrm{\overset{\displaystyle R}{\underset{\displaystyle COOH}{C}}=NH} \xrightarrow{+\,H_2O} \mathrm{\overset{\displaystyle R}{\underset{\displaystyle COOH}{C}}=O} + \mathrm{NH_3}$$

Knoop [8], in his first hypothesis, considered the formation of an amino acid as the reversal of this scheme and due to an enzymatic reduction of the non-enzymatically formed imino acid. In favour of this first hypothesis, Knoop and Oesterlin [11] referred to a chemical model in which they added reducing agents (for instance cysteine) to a mixture of keto acids and NH_3 and obtained amino acids.

2. Acetylating amination of keto acids

Knoop proposed a second hypothesis derived from his observations on the catabolism of γ-phenyl-α-aminobutyric acid. He gave to a dog, in the course of three days, 18 g of the inactive amino acid and he recovered in the urine the following substances: *laevo*-phenyl-α-aminobutyric acid, the acetyl derivative of the *dextro*-phenyl-α-aminobutyric acid, *dextro*-phenyl-α-hydroxybutyric acid, hippuric acid and a residue giving the reaction of an α-ketonic acid.

As stated by Dakin [12] referring to these observations:

"The catabolism of the amino acid evidently was similar to that of phenyl-α-aminoacetic acid, and it appeared likely that the hydroxy acid was formed by the asymmetric reduction of the ketonic acid. This supposition was subsequently confirmed by actually administering the sodium salt of the ketonic acid (12 g) to a dog and obtaining the dextro-hydroxy acid (2.5 g) from the urine. But in addition 0.44 g of the acetyl derivative of dextro-phenyl-α-aminobutyric acid was obtained i.e. a synthesis of the amino acid from the ketonic acid involving reduction had taken place.

"The various changes may be represented as follows:

ketonic acid involving reduction had taken place.

"The various changes may be represented as follows:

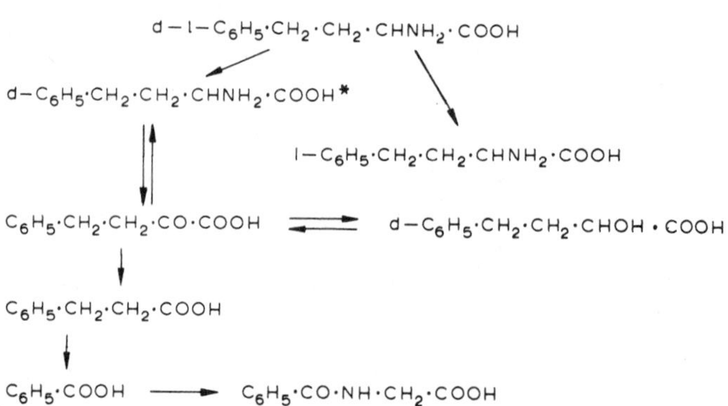

* Excreted in part in the form of its acetyl derivative.

"Since administration of γ-phenyl-α-hydroxybutyric also gave rise to the excretion of a small amount of the dextro-acetyl derivative of γ-phenyl-α-aminobutyric acid, it appears likely that the hydroxy acid may be oxidized in the body to the ketonic acid (a reversible reaction) and so in turn may yield the amino acid." (p. 55).

Knoop explained the finding of hippuric acid in the urine by a further oxidation of the ketonic acid to yield phenylpropionic acid which then underwent β-oxidation, yielding benzoic acid which is excreted as hippuric acid. As stated above after feeding γ-phenyl-α-aminobutyric acid, Knoop found in the urine among other compounds the acetyl derivative of phenyl-α-aminobutyric acid, and he observed an increase in this excretion if he gave the animal pyruvic acid. Knoop considered that the acetyl group could be derived from the dehydrogenation of a molecule of pyruvic acid and that the hydrogen becoming available through this process could be used for the reductive process of fixation of the ammonia to the phenyl-keto-butyric acid. A model was available in the chemical studies of the formation of alanine from pyruvate and NH_3.

Knoop proposed a scheme of the reaction which he designated as "acetylating amination"

$$\begin{array}{c} R \\ | \\ C{=}O \\ | \\ COOH \end{array} \xrightarrow{+NH_3} \begin{array}{c} R \\ | \\ C{=}NH \\ | \\ COOH \end{array} + \begin{array}{c} CH_3 \\ | \\ C{=}O \\ | \\ COOH \end{array} \longrightarrow \begin{array}{c} R \qquad\qquad CH_3 \\ | \qquad\qquad\quad | \\ C{\equiv}N{-}C \\ | \qquad\qquad\quad | \\ COOH \quad H\,OCO \end{array} \longrightarrow \begin{array}{c} R \qquad CH_3 \\ | \qquad\quad | \\ HC{-}NH{\cdot}CO \\ | \\ COOH \end{array} + CO_2$$

If this process starts from natural keto acids, it is to be expected that the acetyl amino acid will rapidly go into amino acid. Therefore, in such event, the excretion of the acetyl derivative in the urine should not be expected, as it is with phenyl derivatives.

The hypothesis, formulated by Knoop, of the "acetylating amination" was reinforced by observations made by Du Vigneaud and Irish [13]. Krebs [14] also interpreted the formation of amino acids from pyruvate in the presence of liver slices (Neber [5]) as an "acetylating amination". According to Krebs, in such a case pyruvate plays the part of a specific hydrogen donor for the imino acid, which was considered as confirmed by the observation that pyruvate cannot be replaced by other keto acids, for instance by ketoglutaric acid.

References p. 333

3. Transamination from glutamate

With respect to the enzymatic aspect of the process of amino acid biosynthesis from keto acids, there first came into consideration a reversibility of the reaction catalysed by glutamate dehydrogenase. The property of glutamic acid as hydrogen donor has been repeatedly stressed (Thunberg [15], Quastel [16]).

It was von Euler and his collaborators [17] who described the reversibility of the reaction of glutamic acid dehydrogenation in the presence of animal glutamate dehydrogenase and realized an enzymatic synthesis (a hydrogenating amination) of glutamic acid from ketoglutaric acid and ammonia. As the enzymatic system is particularly concentrated in the liver it was considered by von Euler, Adler, Günther and Das [18] that the liver is the main seat of glutamate biosynthesis.

Referring to the discovery of transamination by Braunstein and Kritzmann (see Part III of this *History*), von Euler and his colleagues suggested that whereas glutamic acid is biosynthesized through reducing amination, the formation of the other amino acids may result from processes of transamination to the corresponding keto acids. As ketoglutaric acid is a normal intermediary product of the catabolism of carbohydrates in cells, it provided a key for the synthesis of amino acid and a way of introduction of NH_3, which after being attached on glutamic acid, from this situation, could be transferred by transamination to different keto acids.

$$NH_3 \xrightarrow{\text{+ketoglutaric acid}} \text{glutamic acid} \xrightarrow{+\alpha\text{-keto acids}} \text{amino acids}\ .$$

4. Von Euler's scheme

Von Euler et al. [18] considered the possibility of other pathways playing a role in the biosynthesis of amino acids, one of those they recognized being the "acetylating amination" described above. As the simplest example of this process they considered the formation of alanine from two molecules of pyruvic acid and NH_3. But they noted that, as already stressed by Knoop and later by Du Vigneaud and Irish, the reaction can take place between 1 molecule each of pyruvic acid, NH_3 and a second keto acid. It was therefore suggested that during a biosynthesis of amino acids in the course of the perfusion of an isolated liver with a keto acid and ammonia, the reduction by the pyruvate formed in the liver takes place in the form of acetylating amination. There is a third possibility considered by von Euler: the possible reversibility of the action of amino acid dehydrogenase ("oxido-desaminase") of Krebs. Von Euler proposed a scheme of amino acid metabolism, including biosynthesis, the background of which is to be found in the studies on dehydrogenations performed at the time, and in which the core of the interpretation lies in the reducing amination of keto-glutaric acid and the transamination from the resulting glutamic acid. This pathway von Euler took as the central point of the scheme, because, at the time of his writing, it was the best known ("weil dieser Weg der gegenwärtig am besten bekannte ist").

5. Proline, glutamic acid and glutamine

Von Euler's scheme shows that what he considered as the key substance of the system of ammonia fixation by keto acids, the keto-glutaric acid, derived, as had been shown by Martius (see Part III) from citric acid and was in close relation to the catabolic pathway of carbohydrates. The connection with oxyglutaric acid is accomplished, according to Thunberg [15] and to Weil-Malherbe [19], through oxyglutaric acid dehydrogenase, in reactions whose reversibility was accepted by these authors. The close relationship between proline and glutamic acid, mentioned in the scheme, had been pointed to by Weil-Malherbe and Krebs [20], in the laboratory of Hopkins. From their results, these authors formulated the metabolism of proline in the kidney of rabbit and guinea pig as follows

$$\text{proline} + O_2 \longrightarrow \text{glutamic acid} \begin{cases} \nearrow + NH_3 \longrightarrow \text{glutamine} \\ \searrow + \tfrac{1}{2}O_2 \longrightarrow \alpha\text{-ketoglutaric acid} + \text{ammonia} \end{cases}$$

In a previous study, Krebs [21] had shown that, in the kidney, ammonium glutamate was converted into glutamine. Trying other amino acids, Krebs recognized that only proline and hydroxy-proline behaved similarly to glutamic acid.

References p. 333

To quote Weil-Malherbe and Krebs [20]:

"If proline or hydroxyproline and ammonium salts are added to kidney, ammonia disappears and the ammonia which has disappeared is found in the solution as amide-nitrogen. The rate of amide-nitrogen formation is smaller in the presence of proline or hydroxyproline than it is in the presence of glutamic acid. With proline the rate of amide-nitrogen formation is 20—30%, whilst with hydroxyproline it is 5—10%, of the rate obtained with *l* — (+) — glutamic acid. When proline or hydroxyproline (without ammonia) is added to kidney slices, less ammonia is formed than in their absence; instead of ammonia amide-nitrogen is found in the solution."

Krebs recognized that "the intermediate stages between proline and glutamic acid are obscure". Weil-Malherbe and Krebs [20] found that in kidney slices, in contrast with glutamic acid, proline, hydroxyproline and pyrrolidone carboxylic acid had no effect on oxygen uptake, ammonia consumption or production of amino acid or amide nitrogen.

If the background of the views of H. von Euler can be recognized in the dehydrogenation studies which followed on the work of Wieland and of Thunberg, the contribution of Krebs and Weil-Malherbe was derived from the report of Krebs in 1935 that, as recalled above, the disappearance of ammonia from slices of guinea pig or rabbit kidney was augmented by L-glutamate and that this disappearance was associated with the formation of acid-labile ammonia (100°C, 5% H_2SO_4, 5 min).

6. Biosynthesis of glycine

That glycine can be biosynthesized by mammals has been known since the beginning of this century.

Feeding benzoate over long periods to animals, Wiechowski [22] and Magnus-Levy [23] observed an excretion of hippuric acid higher than could be accounted for by either the glycine present in the body proteins or that ingested. Using pure amino acids in the diet, McCoy and Rose [24] and Rose, Burr and Sallach [25] demonstrated that neither glycine nor serine had to be given to young rats in order to ensure a good rate of growth. While glycine is a non-essential amino acid for mammals, this is not so with birds. Almquist, Stokstad, Mecchi and Manning [26], as well as Almquist and Gran [27] showed that growth can be greatly improved by the addition of glycine to the diet of hens deficient in this amino acid.

For optimal growth, the turkey also requires glycine, as was shown by Jukes, Stokstad and Belt [28] although the amounts required appeared smaller than in the hen (Kratzer and Williams [29]).

The concept, now current, according to which the major pathway of glycine biosynthesis is via the decomposition of serine to glycine and methylene-FH_4 was not the first of the theories proposed for the biosynthesis of glycine.

In 1909, Neubauer [30] observed that phenylglycine gave rise to phenylglyoxylic acid and he proposed that glycine was formed by a reverse process, the amination of glyoxylic acid. A finding of Ratner, Nocito and Green [31] gave more credit to the theory. These authors showed that there is, in liver and kidney, an enzyme that catalyses the transformation of glycine into glyoxylic acid and ammonia. Nevertheless, perfusion or feeding experiments failed to substantiate the theory (Hass [32]; Sassa [33]).

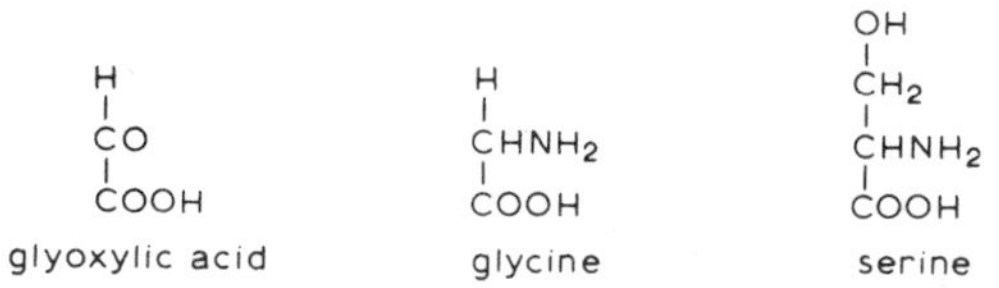

H–CO–COOH	H–CHNH$_2$–COOH	OH–CH$_2$–CHNH$_2$–COOH
glyoxylic acid	glycine	serine

Knoop [34] (in 1914) considered that at least one amino acid, glycine, was biosynthesized in the animal body. He accepted the concept proposed by Neubauer, that from the amino acids, the catabolism liberated the next lower fatty acid. In this process, the nitrogen was also liberated and the resulting compounds were of an entirely different nature from the original amino acid. Knoop considered serine as a possible precursor of glycine. He started from the existence of an α-amino and a β-oxy group in serine and considered the possibility of the attack on any of these two:

"Im ersteren Falle müsste sich Oxybrenztrauben-säure bilden, im anderen Falle würde Aminomalonsäure entstehen, die in vitro leicht CO_2 abspaltet und also Glykokoll liefern könnte."

As direct determination of the expected products was out of the question in experiments on whole organisms, Knoop, according to

References p. 333

his methodological tendencies, relied on an indirect procedure. He found that β-phenyl-serine, when fed to a dog, yielded hippuric acid, and concluded that α-amino-β-hydroxy acids undergo β oxidation to yield glycine. But, as noted by Shemin [35],

> "...the theory actually sheds no light on the nature of the first 2-carbon fragment that is split off, for benzoic acid should be formed from these two odd numbered acids, which in turn would give rise to hippuric acid, regardless of the nature of the 2-carbon fragment."

Whether glycollic acid can serve as a precursor of glycine was considered by several authors. Griffith [36] found that in rats given, in addition to an adequate diet, toxic quantities of sodium benzoate, the toxicity of this compound was decreased as well, and as effectively, by glycollic acid as by glycine. Milhorat and Toscani [37] studied the effect of glycollic acid upon the excretion of creatine in patients with progressive muscular dystrophy. As it had been shown (references in Milhorat and Toscani [37]) that the administration of glycine in progressive muscular dystrophy is followed by an increased excretion of urinary creatine, the authors considered that, if glycollic acid can be converted into glycine in the body, its ingestion by patients should increase the output of creatine in a manner similar to that following the ingestion of glycine. But the authors observed that the effect of glycollic acid was insignificant compared with that of glycine.

REFERENCES

1 E. Knoop and E. Kertess, Z. Physiol. Chem., 71 (1911) 252.
2 F. Knoop and J. Garcia Blanco, Z. Physiol. Chem., 146 (1925) 267.
3 H. Embden and E. Schmitz, Biochem. Z., 29 (1910) 423; 38 (1912).
4 H.D. Dakin and H.W. Dudley, J. Biol. Chem., 18 (1914) 29.
5 M. Neber, Z. Physiol. Chem., 234 (1935) 83.
6 O. Neubauer and W. Gross, Z. Physiol. Chem., 67 (1910) 219.
7 O. Neubauer and K. Kromherz, Z. Physiol. Chem., 70 (1911) 326.
8 F. Knoop, Z. Physiol. Chem., 167 (1910) 489.
9 H.A. Krebs, Z. Physiol. Chem., 217 (1933) 191; 218 (1933) 157.
10 H.A. Krebs, Biochem. J., 29 (1935) 1620.
11 F. Knoop and H. Oesterlin, Z. Physiol. Chem., 148 (1925) 294.
12 H.D. Dakin, Oxidations and Reductions in the Animal Body, London, 1912.
13 V. du Vigneaud and O.J. Irish, J. Biol. Chem., 122 (1938) 349.
14 H.A. Krebs, Ann. Rev. Biochem., 5 (1936) 247.
15 T. Thunberg, Skand. Arch. Physiol., 40 (1920) 1.
16 J.H. Quastel and A.H.M. Wheatley, Biochem. J., 26 (1932) 725.
17 E. Adler, N.B. Das, H. von Euler and Heyman, C.R. Trav. Lab. Carlsberg, Ser. Chim., 22 (1938) 15.
18 H. von Euler, E. Adler, G. Günther and N.B. Das, Z. Physiol. Chem., 254 (1938) 61.
19 H. Weil-Malherbe, Biochem. J., 31 (1937) 2080.
20 H. Weil-Malherbe and H.A. Krebs, Biochem. J., 29 (1935) 2077.
21 H.A. Krebs, Biochem. J., 29 (1935) 1951.
22 W. Wiechowski, Beitr. Chem. Physiol. Pathol., 7 (1906) 204.
23 A. Magnus-Levy, Biochem. Z., 6 (1907) 523.
24 R.H. McCoy and W.C. Rose, J. Biol. Chem., 117 (1937) 581.
25 W.C. Rose, W.W. Burr and H.J. Sallach, J. Biol. Chem., 194 (1952) 321.
26 H.J. Almquist, E.L. Stokstad, E. Mecchi and P.D.V. Manning, J. Biol. Chem., 134 (1940) 213.
27 H.J. Almquist and C.R. Gran, J. Nutr., 28 (1944) 325.
28 T.H. Jukes, E.L. Stokstad and M. Belt, J. Nutr., 33 (1947) 1.
29 F.H. Kratzer and D. Williams, J. Nutr., 35 (1948) 315.
30 O. Neubauer, Dtsch. Arch. Klin. Med., 95 (1909) 211, cited after Shemin [35].
31 S. Ratner, V. Nocito and D.E. Green, J. Biol. Chem., 152 (1944) 119.
32 G. Haas, Biochem. Z., 46 (1912) 296.
33 R. Sassa, Biochem. Z., 59 (1914) 353.
34 F. Knoop, Z. Physiol. Chem., 89 (1914) 151.
35 D. Shemin, J. Biol. Chem., 162 (1946) 297.
36 W.H. Griffith, J. Biol. Chem., 100 (1933) 1.
37 A.T. Milhorat and V. Toscani, J. Biol. Chem., 114 (1936) 461.

Subject Index

Name Index

NOTES ADDED IN PROOF

(p. 153)

In a letter to the author, Chibnall has pointed out that the switch to history after a career in science is not as unique as might be supposed:

"The late Dr. Cr. Herbert Fowler (1861—1940) was a distinguished professor of zoology at University College, London, until 1912, when he retired to a village in Bedfordshire and in that same year founded as a publishing body the Bedfordshire Historical Record Society, to which he contributed many papers and volumes dealing with the medieval history of the country".

(p. 322)

Another example was given by the biosynthesis of acetylcholine which is not the reversal of the hydrolysis. As shown by Nachmansohn, Cox, Coates and Machado [52] the hydrolysis is brought about by choline esterase, which is inhibited by eserine, while the biosynthesis is achieved by choline acetylase which is not affected by eserine. Krebs [53], in 1946, has called the attention to the danger of considering as a reversible enzyme action what is in reality an irreversible cycle.